N.F. Dupuis

Elementary synthetic geometry of the point, line and circle in the plane

N.F. Dupuis

Elementary synthetic geometry of the point, line and circle in the plane

ISBN/EAN: 9783742892355

Manufactured in Europe, USA, Canada, Australia, Japa

Cover: Foto ©berggeist007 / pixelio.de

Manufactured and distributed by brebook publishing software (www.brebook.com)

N.F. Dupuis

Elementary synthetic geometry of the point, line and circle in the plane

ELEMENTARY
SYNTHETIC GEOMETRY

OF THE

POINT, LINE AND CIRCLE IN

THE PLANE

BY

N. F. DUPUIS, M.A., F.R.S.C.

PROFESSOR OF PURE MATHEMATICS IN THE UNIVERSITY OF QUEEN'S
COLLEGE, KINGSTON, CANADA.

London:
MACMILLAN AND CO.
AND NEW YORK.
1889.

[All rights reserved]

PREFACE.

The present work is a result of the Author's experience in teaching Geometry to Junior Classes in the University for a series of years. It is not an edition of "Euclid's Elements," and has in fact little relation to that celebrated ancient work except in the subject matter.

The work differs also from the majority of modern treatises on Geometry in several respects.

The point, the line, and the curve lying in a common plane are taken as the geometric elements of Plane Geometry, and any one of these or any combination of them is defined as a geometric plane figure. Thus a triangle is not the three-cornered portion of the plane inclosed within its sides, but the combination of the three points and three lines forming what are usually termed its vertices and its sides and sides produced.

This mode of considering geometric figures leads

naturally to the idea of a figure as a locus, and consequently prepares the way for the study of Cartesian Geometry. It requires, however, that a careful distinction be drawn between figures which are capable of superposition and those which are equal merely in area. The properties of congruence and equality are accordingly carefully distinguished.

The principle of motion in the transformation of geometric figures, as recommended by Dr. Sylvester, and as a consequence the principle of continuity are freely employed, and an attempt is made to generalize all theorems which admit of generalization.

An endeavour is made to connect Geometry with Algebraic forms and symbols, (1) by an elementary study of the modes of representing geometric ideas in the symbols of Algebra, and (2) by determining the consequent geometric interpretation which is to be given to each interpretable algebraic form. The use of such forms and symbols not only shortens the statements of geometric relations but also conduces to greater generality.

In dealing with proportion the method of *measures* is employed in preference to that of *multiples* as being equally accurate, easier of comprehension, and more in line with elementary mathematical study. In dealing with ratio I have ventured, when comparing two finite lines, to introduce Hamilton's word *tensor* as seeming to me to express most clearly what is meant.

After treating of proportion I have not hesitated to employ those special ratios known as trigonometric functions in deducing geometric relations.

In the earlier parts of the work Constructive Geometry is separated from Descriptive Geometry, and short descriptions are given of the more important geometric drawing-instruments, having special reference to the geometric principles of their actions.

Parts IV. and V. contain a synthetic treatment of the theories of the mean centre, of inverse figures, of pole and polar, of harmonic division, etc., as applied to the line and circle; and it is believed that a student who becomes acquainted with these geometric extensions in this their simpler form will be greatly assisted in the wider discussion of them in analytical conics. Throughout the whole work modern terminology and modern processes have been used with the greatest freedom, regard being had in all cases to perspicuity.

As is evident from what has been said, the whole intention in preparing the work has been to furnish the student with that kind of geometric knowledge which may enable him to take up most successfully the modern works on Analytical Geometry.

N. F. D.

QUEEN'S COLLEGE,
    KINGSTON, CANADA.

CONTENTS.

PART I.

PAGE

SECTION I.—The Line and Point. SECTION II.—Two Lines—Angles. SECTION III.—Three or more Lines and Determined Points—The Triangle. SECTION IV.—Parallels. SECTION V.—The Circle. SECTION VI.—Constructive Geometry, . . . 1

PART II.

SECTION I.—Comparison of Areas. SECTION II.—Measurement of Lengths and Areas. SECTION III.—Geometric Interpretation of Algebraic Forms. SECTION IV.—Areal Relations—Squares and Rectangles. SECTION V.—Constructive Geometry, . 91

PART III.

SECTION I.—Proportion amongst Line-Segments. SECTION II.—Functions of Angles and their Applications in Geometry, ... 147

PART IV.

SECTION I.—Geometric Extensions. SECTION II.—Centre of Mean Position. SECTION III.—Collinearity and Concurrence. SECTION IV.—Inversion and Inverse Figures. SECTION V.—Pole and Polar. SECTION VI.—The Radical Axis. SECTION VII.—Centres and Axes of Perspective or Similitude, ... 178

PART V.

SECTION I.—Anharmonic Division. SECTION II.—Harmonic Ratio. SECTION III.—Anharmonic Properties. SECTION IV.—Polar Reciprocals and Reciprocation. SECTION V.—Homography and Involution, ... 252

ELEMENTARY SYNTHETIC GEOMETRY.

PART I.

GENERAL CONSIDERATIONS.

1°. A statement which explains the sense in which some word or phrase is employed is a *definition*.

A definition may select some one meaning out of several attached to a common word, or it may introduce some technical term to be used in a particular sense.

Some terms, such as space, straight, direction, etc., which express elementary ideas cannot be defined.

2°. *Def.*—A *Theorem* is the formal statement of some mathematical relation.

A theorem may be stated for the purpose of being subsequently proved, or it may be deduced from some previous course of reasoning.

In the former case it is called a *Proposition*, that is, something proposed, and consists of (*a*) the statement or enunciation of the theorem, and (*b*) the argument or proof. The purpose of the argument is to show that the truth of the theorem depends upon that of some preceding theorem whose truth has already been established or admitted.

Ex. "The sum of two odd numbers is an even number" is a theorem.

3°. A theorem so elementary as to be generally accepted as true without any formal proof, is an *axiom*.

Mathematical axioms are general or particular, that is, they apply to the whole science of mathematics, or have special applications to some department.

The principal general axioms are :—

    i. The whole is equal to the sum of all its parts, and therefore greater than any one of its parts.

    ii. Things equal to the same thing are equal to one another.

    iii. If equals be added to equals the sums are equal.

    iv. If equals be taken from equals the remainders are equal.

    v. If equals be added to or taken from unequals the results are unequal.

    vi. If unequals be taken from equals the remainders are unequal.

    vii. Equal multiples of equals are equal; so also equal submultiples of equals are equal.

The axioms which belong particularly to geometry will occur in the sequel.

4°. The statement of any theorem may be put into the hypothetical form, of which the type is—

    If A is B then C is D.

The first part "if A is B" is called the *hypothesis*, and the second part "then C is D" is the *conclusion*.

Ex. The theorem "The product of two odd numbers is an odd number" can be arranged thus :—

    Hyp. If two numbers are each an odd number.
    Concl. Then their product is an odd number.

5°. The statement "If A is B then C is D" may be immediately put into the form—

    If C is not D then A is not B,

which is called the contrapositive of the former.

The truth of a theorem establishes the truth of its contrapositive, and *vice versa*, and hence if either is proved the other is proved also.

6°. Two theorems are converse to one another when the hypothesis and conclusion of the one are respectively the conclusion and hypothesis of the other.

Ex. If an animal is a horse it has four legs.

Converse. If an animal has four legs it is a horse.

As is readily seen from the foregoing example, the truth of a theorem does not necessarily establish the truth of its converse, and hence a theorem and its converse have in general to be proved separately. But on account of the peculiar relation existing between the two, a relation exists also between the modes of proof for the two. These are known as the *direct* and *indirect* modes of proof. And if any theorem which admits of a converse can be proved directly its converse can usually be proved indirectly. Examples will occur hereafter.

7°. Many geometric theorems are so connected with their converses that the truth of the theorems establishes that of the converses, and *vice versa*.

The necessary connection is expressed in the *Rule of Identity*, its statement being :—

If there is but one X and one Y, and if it is proved that X is Y, then it follows that Y is X.

Where X and Y stand for phrases such as may form the hypotheses or conclusions of theorems, and the "is" between them is to be variously interpreted as "equal to," "corresponds to," etc.

Ex. Of two sides of a triangle only one can be the greater, and of the two angles opposite these sides only one can be the greater. Then, if it is proved that the greater side is opposite the greater angle it follows that the greater angle is opposite the greater side.

4 SYNTHETIC GEOMETRY.

In this example there is but one X (the greater side) and one Y (the greater angle), and as X is (corresponds to or is opposite) Y, therefore Y is (corresponds to or is opposite) X.

8°. A *Corollary* is a theorem deduced from some other theorem, usually by some qualification or restriction, and occasionally by some amplification of the hypothesis. Or a corollary may be derived directly from an axiom or from a definition.

As a matter of course no sharp distinction can be drawn between theorems and corollaries.

Ex. From the theorem, " The product of two odd numbers is an odd number," by making the two numbers equal we obtain as a corollary, " The square of an odd number is an odd number."

EXERCISES.

State the contrapositives and the converses of the following theorems :—

1. The sum of two odd numbers is an even number.
2. A diameter is the longest chord in a circle.
3. Parallel lines never meet.
4. Every point equidistant from the end-points of a line-segment is on the right bisector of that segment.

SECTION I.

THE LINE AND POINT.

9°. Space may be defined to be that which admits of length or distance in every direction; so that length and direction are fundamental ideas in studying the geometric properties of space.

THE LINE AND POINT.

Every material object exists in, and is surrounded by space. The limit which separates a material object from the space which surrounds it, or which separates the space occupied by the object from the space not occupied by it, is a *surface*.

The surface of a black-board is the limit which separates the black-board from the space lying without it. This surface can have no thickness, as in such a case it would include a part of the board or of the space without or of both, and would not be the dividing limit.

10°. A flat surface, as that of a black-board, is a plane surface, or a *Plane*.

Pictures of geometric relations drawn on a plane surface as that of a black-board are usually called *Plane Geometric Figures*, because these figures lie in or on a plane.

Some such figures are known to every person under such names as "triangle," "square," "circle," etc.

11°. That part of mathematics which treats of the properties and relations of plane geometric figures is *Plane Geometry*. Such is the subject of this work.

The plane upon which the figures are supposed to lie will be referred to as *the plane*, and unless otherwise stated all figures will be supposed to lie in or on the same plane.

12°. *The Line.* When the crayon is drawn along the black-board it leaves a visible mark. This mark has breadth and occupies some of the surface upon which it is drawn, and by way of distinction is called a *physical line*. By continually diminishing the breadth of the physical line we make it approximate to the geometric line. Hence we may consider the geometric line as being the limit towards which a physical line approaches as its breadth is continually diminished. We may consequently consider a geometric line as *length* abstracted from every other consideration.

This theoretic relation of a geometric line to a physical one is of some importance, as whatever is true for the physical line, independently of its breadth, is true for the geometric line. And hence arguments in regard to geometric lines may be replaced by arguments in regard to physical lines, if from such arguments we exclude everything that would involve the idea of breadth.

The diagrams employed to direct and assist us in geometric investigations are formed of physical lines, but they may equally well be supposed to be formed of threads, wires or light rods, if we do not involve in our arguments any idea of the breadth or thickness of the lines, threads, wires or rods employed.

In the practical applications of Geometry the diagrams frequently become material or represent material objects. Thus in Mechanics we consider such things as levers, wedges, wheels, cords, etc., and our diagrams become representations of these things.

A pulley or wheel becomes a circle, its arms become radii of the circle, and its centre the centre of the circle; stretched cords become straight lines, etc.

13°. *The Point.* A point marks position, but has no size. The intersection of one line by another gives a point, called the point of intersection.

If the lines are physical, the point is physical and has some size, but when the lines are geometric the point is also geometric.

14°. *Straight Line.* For want of a better definition we may say that a *straight line* is one of which every part has the same direction. For every part of a line must have some direction, and when this direction is common to all the parts of the line, the line is straight.

The word "direction" is not in itself definable, and when applied to a line in the absolute it is not intelligible. But every person knows what is meant by such expressions as "the same direction," "opposite direction," etc., for these express relations between directions, and such relations are as readily comprehended as relations between lengths or other magnitudes.

The most prominent property, and in fact the distinctive property of a *straight* line, is the absolute sameness which characterizes all its parts, so that two portions of the same straight line can differ from one another in no respect except in length.

Def.—A plane figure made up of straight lines only is called a *rectilinear* figure.

15°. A *Curve* is a line of which no part is straight; or a curve is a line of which no two adjacent parts have the same direction.

The most common example of a curve is a circle or portion of a circle.

Henceforward, the word "line," unless otherwise qualified, will mean a *straight* line.

16°. The "rule" or "straight-edge" is a strip of wood, metal, or other solid with one edge made *straight*. Its common use is to guide the pen or pencil in drawing lines in Practical Geometry.

17°. A *Plane* is a surface such that the line joining any two arbitrary points in it coincides wholly with the surface.

The planarity of a surface may be tested by applying the rule to it. If the rule touches the surface at some points and not at others the surface is not a plane. But if the rule touches the surface throughout its whole length, and in every position and direction in which it can be applied, the surface is a plane.

The most accurately plane artificial surface known is probably that of a well-formed plane mirror. Examination of the images of objects as seen in such mirrors is capable of detecting variations from the plane, so minute as to escape all other tests.

18°. A surface which is not plane, and which is not composed of planes, is a curved surface. Such is the surface of a sphere, or cylinder.

19°. The point, the line, the curve, the plane and the curved surface are the elements which go to make up geometric figures.

Where a single plane is the only surface concerned, the point and line lie in it and form a plane figure. But where more than one plane is concerned, or where a curved surface is concerned, the figure occupies space, as a cube or a sphere, and is called a *spatial* figure or a *solid*.

The study of spatial figures constitutes Solid Geometry, or the Geometry of Space, as distinguished from Plane Geometry.

20°. *Given Point and Line.* A point or line is said to be *given* when we are made to know enough about it to enable us to distinguish it from every other point or line; and the data which give a point or line are commonly said to *determine* it.

A similar nomenclature applies to other geometric elements.

The statement that a point or line lies in a plane does not give it, but a point or line placed in the plane for future reference is considered as being given. Such a point is usually called an *origin*, and such a line a *datum line*, an *initial line*, a *prime vector*, etc.

21°. *Def.* 1.—A line considered merely as a geometric element, and without any limitations, is an *indefinite* line.

2.—A limited portion of a line, especially when any reference is had to its length, is a *finite* line, or a *line-segment*, or simply a *segment*.

That absolute sameness (14°) which characterizes every part of a line leads directly to the following conclusions :—
(1) No distinction can be made between any two segments of the same line equal in length, except that of position in the line.
(2) A line cannot return into, or cross itself.
(3) A line is not necessarily limited in length, and hence, in imagination, we may follow a line as far as we please without coming to any necessary termination.

This property is conveniently expressed by saying that a line *extends to infinity*.

3.—The hypothetical end-points of any indefinite line are said to be points at infinity. All other points are *finite points*.

22°. *Notation.* A point is denoted by a single letter wherever practicable, as "the point A."

An indefinite line is also denoted by a single letter as "the line L," but in this case the letter has no reference to any point.

A segment is denoted by naming its end points, as the "segment AB," where A and B are the *end points*. This is a *biliteral*, or two-letter notation.

A segment is also denoted by a single letter, when the limits of its length are supposed to be known, as the "segment *a*." This is a *uniliteral*, or one-letter notation.

The term "segment" involves the notion of some finite length. When length is not under consideration, the term "line" is preferred.

Thus the "line AB" is the indefinite line having A and B as two points upon it. But the "segment AB" is that portion of the line which lies between A and B.

23°. In dealing with a line-segment, we frequently have to consider other portions of the indefinite line of which the segment is a part.

As an example, let it be required to divide the segment AB into two parts whereof one shall be twice as long as the other. To do this we put C in such a position that it may be twice as far from one of the end-points of the segment, A say, as it is from the other, B. But on the indefinite line through A and B we may place C' so as to be twice as far from A as from B. So that we have two points, C and C', both satisfying the condition of being twice as far from A as from B.

Evidently, the point C' does not divide the segment AB in the sense commonly attached to the word *divide*. But on account of the similar relations held by C and C' to the end-points of the segment, it is convenient and advantageous to consider both points as dividing the segment AB.

When thus considered, C is said to divide the segment *internally* and C' to divide it *externally* in the same manner.

24°. *Axiom.*—Through a given point only one line can pass in a given direction.

Let A be the given point, and let the segment AP mark the given direction. Then, of all the lines that can pass through the point A, only one can have the direction AP, and this one must lie along and coincide with AP so as to form with it virtually but one line.

Cor. 1. A finite point and a direction determine one line.

Cor. 2. Two given finite points determine one line. For, if A and P be the points, the direction AP is given, and hence the line through A and having the direction AP is given.

Cor. 3. Two lines by their intersection determine one finite point. For, if they determined two, they would each pass through the same two points, which, from Cor. 2, is impossible.

Cor. 4. Another statement of Cor. 2 is—Two lines which have two points in common coincide and form virtually but one line.

THE LINE AND POINT.

25°. *Axiom.*—A straight line is the shortest distance between two given points.

Although it is possible to give a reasonable proof of this axiom, no amount of proof could make its truth more apparent.

The following will illustrate the axiom. Assume any two points on a thread taken as a physical line. By separating these as far as possible, the thread takes the form which we call straight, or tends to take that form. Therefore a straight finite line has its end-points further apart than a curved line of equal length. Or, a less length of line will reach from one given point to another when the line is straight than when it is curved.

Def.—The *distance* between two points is the length of the segment which connects them or has them as end-points.

26°. *Superposition.*—*Comparison of Figures.*—We assume that space is homogeneous, or that all its parts are alike, so that the properties of a geometric figure are independent of its position in space. And hence we assume that a figure may be supposed to be moved from place to place, and to be turned around or over in any way without undergoing any change whatever in its form or properties, or in the relations existing between its several parts.

The imaginary placing of one figure upon another so as to compare the two is called *superposition.* By superposition we are enabled to compare figures as to their equality or inequality. If one figure can be superimposed upon another so as to coincide with the latter in every part, the two figures are necessarily and identically equal, and become virtually one figure by the superposition.

27°. Two line-segments can be compared with respect to length only. Hence a line is called a magnitude of *one dimension.*

Two segments are *equal* when the end-points of one can be

made to coincide with the end-points of the other by superposition.

28°. *Def.*—The *sum* of two segments is that segment which is equal to the two when placed in line with one end-point in each coincident.

Let AB and DE be two segments, and on the line of which AB is a segment let BC be equal to DE. Then AC is the *sum* of AB and DE.

This is expressed symbolically by writing

$$AC = AB + DE,$$

where = denotes equality in length, and + denotes the placing of the segments AB and DE in line so as to have one common point as an end-point for each. The interpretation of the whole is, that AC is equal *in length* to AB and DE together.

29°. *Def.*—The *difference* between two segments is the segment which remains when, from the longer of the segments, a part is taken away equal in length to the shorter.

Thus, if AC and DE be two segments of which AC is the longer, and if BC is equal to DE, then AB is the difference between AC and DE.

This is expressed symbolically by writing

$$AB = AC - DE,$$

which is interpreted as meaning that the segment AB is shorter than AC by the segment DE.

Now this is equivalent to saying that AC is longer than AB by the segment DE, or that AC is equal to the sum of AB and DE.

Hence when we have   AB = AC − DE
we can write   AC = AB + DE.

We thus see that in using these algebraic symbols, =, +, and −, a term, as DE, may be transferred from one side of

THE LINE AND POINT.

the equation to another by changing its sign from + to − or *vice versa*.

Owing to the readiness with which these symbolic expressions can be manipulated, they seem to represent simple algebraic relations, hence beginners are apt to think that the working rules of algebra must apply to them as a matter of necessity. It must be remembered, however, that the formal rules of algebra are founded upon the properties of numbers, and that we should not assume, without examination, that these rules apply without modification to that which is not number.

This subject will be discussed in Part II.

30°. *Def.*—That point, in a line-segment, which is equidistant from the end-points is the *middle* point of the segment.

'It is also called the *internal point of bisection* of the segment, or, when spoken of alone, simply the *point of bisection*.

Exercises.

1. If two segments be in line and have one common endpoint, by what name will you call the distance between their other end-points?
2. Obtain any relation between "the sum and the difference" of two segments and "the relative directions" of the two segments, they being in line.
3. A given line-segment has but one middle point.
4. In Art. 23°, if C becomes the middle point of AB, what becomes of C'?
5. In Art. 30° the internal point of bisection is spoken of. What meaning can you give to the "external point of bisection"?

SECTION II.

RELATIONS OF TWO LINES.—ANGLES.

31°. When two lines have not the same direction they are said to make an *angle* with one another, and an *angle* is a *difference in direction.*

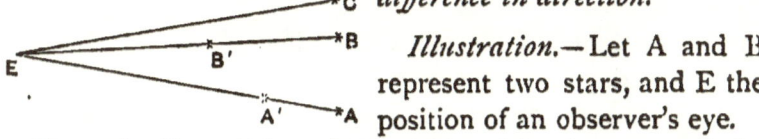

Illustration.—Let A and B represent two stars, and E the position of an observer's eye.

Since the lines EA and EB, which join the eye and the stars, have not the same direction they make an angle with one another at E.

1. If the stars appear to recede from one another, the angle at E becomes greater. Thus, if B moves into the position of C, the angle between EA and EC is greater than the angle between EA and EB.

Similarly, if the stars appear to approach one another, the angle at E becomes smaller; and if the stars become coincident, or situated in the same line through E, the angle at E vanishes.

Hence an angle is capable of continuous increase or diminution, and is therefore a magnitude. And, being magnitudes, angles are capable of being compared with one another as to greatness, and hence, of being measured.

2. If B is moved to B′, any point on EB, and A to A′, any point on EA, the angle at E is not changed. Hence increasing or diminishing one or both of the segments which form an angle does not affect the magnitude of the angle.

Hence, also, there is no community in kind between an angle and a line-segment or a line.

Hence, also, an angle cannot be measured by means of line-segments or lines.

32°. Def.—A line which changes its direction in a plane while passing through a fixed point in the plane is said to *rotate* about the point.

The point about which the rotation takes place is the *pole*, and any segment of the rotating line, having the pole as an end-point, is a *radius vector*.

Let an inextensible thread fixed at O be kept stretched by a pencil at P. Then, when P moves, keeping the thread straight, OP becomes a radius vector rotating about the pole O.

When the vector rotates from direction OP to direction OP' it describes the angle between OP and OP'. Hence we have the following :—

Def. 1.—*The angle between two lines is the rotation necessary to bring one of the lines into the direction of the other.*

The word "rotation," as employed in this definition, means the amount of turning effected, and not the process of turning.

Def. 2.—For convenience the lines OP and OP', which, by their difference in direction form the angle, are called the *arms* of the angle, and the point O where the arms meet is the *vertex*.

Cor. From 31°, 2, an angle does not in any way depend upon the lengths of its arms, but only upon their relative directions.

33°. *Notation of Angles.*—1. The symbol ∠ is used for the word "angle."

2. When two segments meet at a vertex the angle between them may be denoted by a single letter placed at the vertex, as the ∠O ; or by a letter with or without an arch of dots, as ∠β ; or by three letters of which the extreme ones denote points upon the arms of the angle and the middle one denotes the vertex, as ∠AOB.

3. The angle between two lines, when the vertex is not pictured, or not referred to, is expressed by ∠(L. M), or L̂M, where L and M denote the lines in the one-letter notation (22°); or ∠(AB, CD), where AB and CD denote the lines in the two-letter notation.

34°. *Def.*—Two angles are equal when the arms of the one may be made to coincide in direction respectively with the arms of the other; or when the angles are described by the same rotation.

Thus, if, when O′ is placed upon O, and O′A′ is made to lie along OA, O′B′ can also be made to lie along OB, the ∠A′O′B′ is equal to ∠AOB. This equality is symbolized thus:

$$\angle A'O'B' = \angle AOB.$$

Where the sign = is to be interpreted as indicating the possibility of coincidence by superposition.

35°. *Sum and Difference of Angles.*—The *sum* of two angles is the angle described by a radius vector which describes the two angles, or their equals, in succession.

Thus if a radius vector starts from coincidence with OA and rotates into direction OP it describes the ∠AOP. If it next rotates into direction OP′ it describes the ∠POP′. But in its whole rotation it has described the ∠AOP′. Therefore,

$$\angle AOP' = \angle AOP + \angle POP'.$$

Similarly, $\angle AOP = \angle AOP' - \angle POP'$.

Def.—When two angles, as AOP and POP′, have one arm in common lying between the remaining arms, the angles are *adjacent* angles.

36°. *Def.*—A radius vector which starts from any given direction and makes a complete rotation so as to return to its original direction describes a *circumangle*, or *perigon*.

One-half of a circumangle is a *straight* angle, and one-fourth of a circumangle is a *right* angle.

37°. *Theorem*.—If any number of lines meet in a point, the sum of all the adjacent angles formed is a circumangle.

OA, OB, OC, ..., OF are lines meeting in O. Then
$$\angle AOB + \angle BOC + \angle COD + \ldots + \angle FOA$$
$$= \text{a circumangle}.$$

Proof.—A radius vector which starts from coincidence with OA and rotates into the successive directions, OB, OC, ..., OF, OA describes in succession the angles AOB, BOC, ..., EOF, FOA.

But in its complete rotation it describes a circumangle (36°).

∴ $\angle AOB + \angle BOC + \ldots + \angle FOA = $ a circumangle. *q.e.d.*

Cor. The result may be thus stated :—

> The sum of all the adjacent angles about a point in the plane is a circumangle.

38°. *Theorem*.—The sum of all the adjacent angles on one side of a line, and about a point in the line is a straight angle.

O is a point in the line AB; then
$$\angle AOC + \angle COB = \text{a straight angle}.$$

Proof.—Let A and B be any two points in the line, and let the figure formed by AB and OC be revolved about AB without displacing the points A and B, so that OC may come into a position OC'.

Then (24°, Cor. 2) O is not displaced by the revolution,

∴ $\angle AOC = \angle AOC'$, and $\angle BOC = \angle BOC'$;
∴ $\angle AOC + \angle BOC = \angle AOC' + \angle BOC'$,

and since the sum of the four angles is a circumangle (37°), therefore the sum of each pair is a straight angle (36°). *q.e.d.*

Cor. 1. The angle between the opposite directions of a line is a straight angle.

Cor. 2. If a radius vector be rotated until its direction is reversed it describes a straight angle. And conversely, if a radius vector describes a straight angle its original direction is reversed.

Thus, if OA rotates through a straight angle it comes into the direction OB. And conversely, if it rotates from direction OA to direction OB it describes a straight angle.

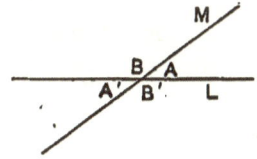

39°. When two lines L and M cut one another four angles are formed about the point of intersection, any one of which may be taken to be the angle between the lines.

These four angles consist of two pairs of *opposite* or *vertical* angles, viz., A, A', and B, B', A being opposite A', and B being opposite B'.

40°. *Theorem.*—The opposite angles of a pair formed by two intersecting lines are equal to one another.

Proof.—	∠A + ∠B = a straight angle	(38°)
and	∠A' + ∠B = a straight angle.	(38°)
∴	∠A = ∠A',	
and	∠B = ∠B'.	*q.e.d.*

Def. 1.—Two angles which together make up a straight angle are *supplementary* to one another, and one is called the *supplement* of the other. Thus, A is the supplement of B', and B of A'.

Cor. If ∠A = ∠B, then ∠A' = ∠B = ∠B', and all four angles are equal, and each is a right angle (36°).

Therefore, if two adjacent angles formed by two intersecting lines are equal to each other, all four of the angles so formed are equal to one another, and each is a right angle.

Def. 2.—When two intersecting lines form a right angle at their point of intersection, they are said to be *perpendicular* to one another, and each is *perpendicular* to the other.

RELATIONS OF TWO LINES.—ANGLES.   19

Perpendicularity is denoted by the symbol ⊥, to be read "perpendicular to" or "is perpendicular to."

A right angle is denoted by the symbol ⌐.

The symbol ⊥ also denotes two right angles or a straight angle.

Def. 3.—When two angles together make up a right angle they are *complementary* to one another, and each is the *complement* of the other.

The right angle is the simplest of all angles, for when two lines form an angle they form four angles equal in opposite pairs. But if any one of these is a right angle, all four are right angles.

Perpendicularity is the most important directional relation in the applications of Geometry.

Def. 4.—An *acute* angle is less than a right angle, and an *obtuse* angle is greater than a right angle, and less than two right angles.

41°. From (36°) we have

$$1 \text{ circumangle} = 2 \text{ straight angles}$$
$$= 4 \text{ right angles.}$$

In estimating an angle numerically it may be expressed in any one of the given units.

If a right angle be taken as the unit, a circumangle is expressed by 4, *i.e.* four right angles, and a straight angle by 2.

Angles less than a right angle may be expressed, approximately at least, by fractions, or as fractional parts of the right angle.

For practical purposes the right angle is divided into 90 equal parts called degrees; each degree is divided into 60 equal parts called minutes; and each minute into 60 equal parts called seconds.

Thus an angle which is one-seventh of a circumangle contains fifty-one degrees, twenty-five minutes, and forty-two seconds and six-sevenths of a second. This is denoted as follows :— $51° \ 25' \ 42\tfrac{6}{7}''$.

42°. *Theorem.*—Through a given point in a line only one perpendicular can be drawn to the line.

The line OC is $\perp$ AB, and OD is any other line through O.

Then OD is not $\perp$ AB.

Proof.—The angles BOC and COA are each right angles (40°, Def. 2).
Therefore BOD is not a right angle, and OD is not $\perp$ AB.
But OD is any line other than OC.
Therefore OC is the only perpendicular. *q.e.d.*

Def.—The perpendicular to a line-segment through its middle point is the *right bisector* of the segment.

Since a segment has but one middle point (30°, Ex. 3), and since but one perpendicular can be drawn to the segment through that point,

∴ *a line-segment has but one right bisector.*

43°. *Def.*—The lines which pass through the vertex of an angle and make equal angles with the arms, are the *bisectors* of the angle. The one which lies within the angle is the *internal* bisector, and the one lying without is the *external* bisector.

Let AOC be a given angle; and let EOF be so drawn that $\angle$AOE $=\angle$EOC.

EF is the internal bisector of the angle AOC.

Also, let GOH be so drawn that $\angle$COG$=\angle$HOA.

HG is the external bisector of the angle AOC.

∵	$\angle$COG$=\angle$HOA	(hyp.)
and	$\angle$HOA$=\angle$GOB,	(40°)
∴	$\angle$COG$=\angle$GOB;	

and the external bisector of AOC is the internal bisector of its supplementary angle, COB, and *vice versa*.

The reason for calling GH a bisector of the angle AOC is given in the definition, viz., GH makes equal angles with the arms. Also, OA and OC are only parts of indefinite lines, whose angle of intersection may be taken as the $\angle$AOC or as the $\angle$COB.

44°. Just as in 23° we found two points which are said to divide the segment in the same manner, so we may find two lines dividing a given angle in the same manner, one dividing it internally, and the other externally.

Thus, if OE is so drawn that the $\angle$AOE is double the $\angle$EOC, some line OG may also be drawn so that the $\angle$AOG is double the $\angle$GOC.

This double relation in the division of a segment or an angle is of the highest importance in Geometry.

45°. *Theorem.*—The bisectors of an angle are perpendicular to one another.

EF and GH are bisectors of the $\angle$AOC;
then EF is $\perp$ GH.

Proof.— $\angle$EOC=$\frac{1}{2}\angle$AOC, $\because$ OE is a bisector;
and $\angle$COG=$\frac{1}{2}\angle$COB, $\because$ OG is a bisector;
adding, $\angle$EOG=$\frac{1}{2}\angle$AOB.
But $\angle$AOB is a straight angle, (38°, Cor. 1)
$\therefore$ $\angle$EOG is a right angle. (36°)

Exercises.

1. Three lines pass through a common point and divide the plane into 6 equal angles. Express the value of each angle in right angles, and in degrees.
2. OA and OB make an angle of 30°, how many degrees are there in the angle made by OA and the external bisector of the angle AOB?

3. What is the supplement of 13° 27′ 42″? What is its complement?
4. Two lines make an angle a with one another, and the bisectors of the angle are drawn, and again the bisectors of the angle between these bisectors. What are the angles between these latter lines and the original ones?
5. The lines L, M intersect at O, and through O, L′ and M′ are drawn ⊥ respectively to L and M. The angle between L′ and M′ is equal to that between L and M.

SECTION III.

THREE OR MORE POINTS AND LINES. THE TRIANGLE.

46°. *Theorem.*—Three points determine at most three lines; and three lines determine at most three points.

Proof 1.—Since (24°, Cor. 2) two points determine one line, three points determine as many lines as we can form groups from three points taken two and two.

Let A, B, C be the points; the groups are AB, BC, and CA.

Therefore three points determine at most three lines.

2.—Since (24°, Cor. 3) two lines determine one point, three lines determine as many points as we can form groups from three lines taken two and two.

But if L, M, N be the lines the groups are LM, MN, and NL.

Therefore three lines determine by their intersections at most three points.

THREE OR MORE POINTS AND LINES.

47°. *Theorem.*—Four points determine at most six lines; and four lines determine at most six points.

Proof.—1. Let A, B, C, D be the four points. The groups of two are AB, AC, AD, BC, BD, and CD; or six in all.

Therefore six lines at most are determined.

2. Let L, M, N, K be the lines. The groups of two that can be made are KL, KM, KN, LM, LN, and MN; or six in all.

Therefore six points of intersection at most are determined.

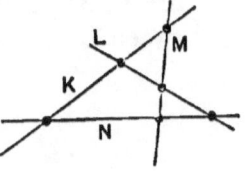

Cor. In the first case the six lines determined pass by threes through the four points. And in the second case the six points determined lie by threes upon the four lines.

This reciprocality of property is very suggestive, and in the higher Geometry is of special importance.

Ex.— Show that 5 points determine at most 10 lines, and 5 lines determine at most 10 points. And that in the first case the lines pass by fours through each point; and in the latter, the points lie by fours on each line.

48°. *Def.*—A *triangle* is the figure formed by three lines and the determined points, or by three points and the determined lines.

The points are the *vertices* of the triangle, and the line-segments which have the points as end-points are the *sides*.

The remaining portions of the determined lines are usually spoken of as the "sides produced." But in many cases generality requires us to extend the term "side" to the whole line.

Thus, the points A, B, C are the vertices of the triangle ABC.

The segments AB, BC, CA are the sides. The portions AE, BF, CD, etc., extending outwards as far as required, are the sides produced.

The triangle is distinctive in being the rectilinear figure for which a given number of lines determines the same number of points, or *vice versa*.

Hence when the three points, forming the vertices, are given, or when the three lines or line-segments forming the sides are given, the triangle is completely given.

This is not the case with a rectilinear figure having any number of vertices other than three.

If the vertices be four in number, with the restriction that each vertex is determined by the intersection of two sides, any one of the figures in the margin will satisfy the conditions.

Hence the giving of the four vertices of such a figure is not sufficient to completely determine the figure.

49°. *Def.*—1. The angles ABC, BCA, CAB are the *internal* angles of the triangle, or simply the *angles* of the triangle.

2. The angle DCB, and others of like kind, are *external* angles of the triangle.

3. In relation to the external angle DCB, the angle BCA is the *adjacent internal angle*, while the angles CAB and ABC are *opposite internal angles*.

4. Any side of a triangle may be taken as its *base*, and then the angles at the extremities of the base are its *basal* angles, and the angle opposite the base is the *vertical* angle. The vertex of the vertical angle is the *vertex* of the triangle when spoken of in relation to the base.

50°. *Notation.*—The symbol △ is commonly used for the word triangle. In certain cases, which are always readily apprehended, it denotes the area of the triangle.

The angles of the triangle are denoted usually by the capital letters A, B, C, and the sides opposite by the corresponding small letters *a*, *b*, *c*.

51°. *Def.*—When two figures compared by superposition coincide in all their parts and become virtually but one figure they are said to be *congruent*.

Congruent figures are distinguishable from one another only by their position in space and are said to be identically equal.

Congruence is denoted by the algebraic symbol of identity, ≡; and this symbol placed between two figures capable of congruence denotes that the figures are congruent.

Closed figures, like triangles, admit of comparison in two ways. The first is as to their capability of perfect coincidence; when this is satisfied the figures are congruent. The second is as to the magnitude or extent of the portions of the plane enclosed by the figures. Equality in this respect is expressed by saying that the figures are *equal*.

When only one kind of comparison is possible, as is the case with line-segments and angles, the word *equal* is used.

CONGRUENCE AMONGST TRIANGLES.

52°. *Theorem.*—Two triangles are congruent when two sides and the included angle in the one are respectively equal to two sides and the included angle in the other.

If AB = A'B'
    BC = B'C' } the triangles are congruent.
and ∠B = ∠B'

Proof.—Place △ABC on △A'B'C' so that B coincides with B', and BA lies along B'A'.

∵   ∠B = ∠B', BC lies along B'C',   (34°)
and ∵   AB = A'B' and BC = B'C';
∴   A coincides with A' and C with C',   (27°)

and ∴ AC lies along A′C′; (24°, Cor. 2)
and the △s coinciding in all their parts are congruent. (51°)
q.e.d.

Cor. Since two congruent triangles can be made to coincide in *all* their parts, therefore—

When two triangles have two sides and the included angle in the one respectively equal to two sides and the included angle in the other, *all* the parts in the one are respectively equal to the *corresponding* parts in the other.

53°. *Theorem*.—Every point upon the right bisector of a segment is equidistant from the end-points of the segment.

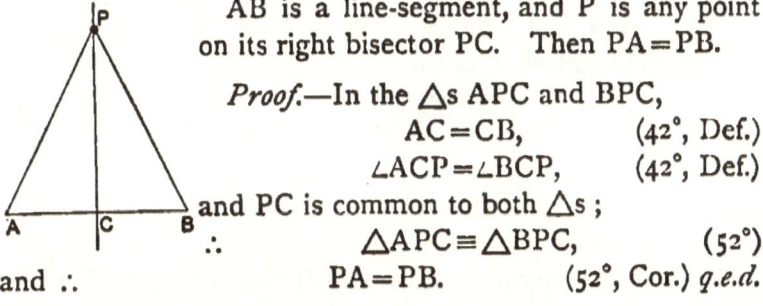

AB is a line-segment, and P is any point on its right bisector PC. Then PA = PB.

Proof.—In the △s APC and BPC,
  AC = CB,   (42°, Def.)
  ∠ACP = ∠BCP, (42°, Def.)
and PC is common to both △s;
  ∴ △APC ≡ △BPC, (52°)
and ∴ PA = PB. (52°, Cor.) *q.e.d.*

Def. 1.—A triangle which has two sides equal to one another is an *isosceles* triangle.

Thus the triangle APB is *isosceles*.

The side AB, which is not one of the equal sides, is called the *base*.

Cor. 1. Since the △APC ≡ △BPC,
  ∴ ∠A = ∠B.
Hence the basal angles of an isosceles triangle are equal to one another.

Cor. 2. From (52°, Cor.), ∠APC = ∠BPC;
Therefore the right bisector of the base of an isosceles triangle is the internal bisector of the vertical angle. And since these two bisectors are one and the same line the converse is true.

Def. 2.—A triangle in which all the sides are equal to one another is an *equilateral* triangle.

Cor. 3. Since an equilateral triangle is isosceles with respect to each side as base, all the angles of an equilateral triangle are equal to one another; or, an equilateral triangle is equiangular.

54°. *Theorem.*—Every point equidistant from the end-points of a line-segment is on the right bisector of that segment. (Converse of 53°.)

PA = PB. Then P is on the right bisector of AB.

Proof.—If P is not on the right bisector of AB, let the right bisector cut AP in Q.

Then $\qquad$ QA = QB, $\qquad$ (53°)
but $\qquad$ PA = PB, $\qquad$ (hyp.)
∴ $\qquad$ QP = PB − QB,
or $\qquad$ PB = QP + QB,
which is not true. $\qquad\qquad\qquad\qquad$ (25°, Ax.)

Therefore the right bisector of AB does not cut AP; and similarly it does not cut BP; therefore it passes through P, or P is on the right bisector. $\qquad$ q.e.d.

This form of proof should be compared with that of Art. 53°, they being the kinds indicated in 6°.

This latter or indirect form is known as proof by *reductio ad absurdum* (leading to an absurdity). In it we prove the conclusion of the theorem to be true by showing that the acceptance of any other conclusion leads us to some relation which is absurd or untrue.

55°. *Def.*—The line-segment from a vertex of a triangle to the middle of the opposite side is a *median* of the triangle.

Cor. 1. Every triangle has three medians.

Cor. 2. The median to the base of an isosceles triangle is

the right bisector of the base, and the internal bisector of the vertical angle. (53°, Cor. 2.)

Cor. 3. The three medians of an equilateral triangle are the three right bisectors of the sides, and the three internal bisectors of the angles.

56°. *Theorem.*— If two angles of a triangle are equal to one another, the triangle is isosceles, and the equal sides are opposite the equal angles. (Converse of 53°, Cor. 1.)

$\angle PAB = \angle PBA$, then $PA = PB$.

Proof.—If P is on the right bisector of AB, $PA = PB$. (53°)

If P is not on the right bisector, let AP cut the right bisector in Q.

Then $QA = QB$, and $\angle QAB = \angle QBA$. (53° and Cor. 1)
But $\angle PBA = \angle QAB$; (hyp.)
∴ $\angle PBA = \angle QBA$,
which is not true unless P and Q coincide.

Therefore if P is not on the right bisector of AB, the $\angle PAB$ cannot be equal to the $\angle PBA$.

But they are equal by hypothesis;
∴ P is on the right bisector,
and $PA = PB$. *q.e.d.*

Cor. If all the angles of a triangle are equal to one another, all the sides are equal to one another.

Or, an equiangular triangle is equilateral.

57°. From 53° and 56° it follows that equality amongst the sides of a triangle is accompanied by equality amongst the angles opposite these sides, and conversely.

Also, that if no two sides of a triangle are equal to one another, then no two angles are equal to one another, and conversely.

THREE OR MORE POINTS AND LINES.   29

Def.—A triangle which has no two sides equal to one another is a *scalene* triangle.

Hence a scalene triangle has no two angles equal to one another.

58°. *Theorem.*—If two triangles have the three sides in the one respectively equal to the three sides in the other the triangles are congruent.

If A'B'=AB ⎫ the △s ABC
   B'C' = BC ⎬ and A'B'C' are
   C'A' = CA ⎭ congruent.

Proof.—Turn the △A'B'C' over and place A' on A, and A'C' along AC, and let B' fall at some point D.

∵   A'C'=AC, C' falls at C,   (27°)
and △ADC is the △A'B'C' in its reversed position.

Since   AB=AD and CB=CD,
A and C are on the right bisector of BD, and AC is the right bisector of BD.   (54°)

∴   ∠BAC=∠DAC ;   (53°, Cor. 2)
and the △s BAC and DAC are congruent.   (52°)
∴   △ABC ≡ △A'B'C'.   q.e.d.

59°. *Theorem.*—If two triangles have two angles and the included side in the one equal respectively to two angles and the included side in the other, the triangles are congruent.

If   ∠A'=∠A ⎫
     ∠C'=∠C ⎬ the △s ABC and A'B'C' are congruent.
and A'C'=AC ⎭

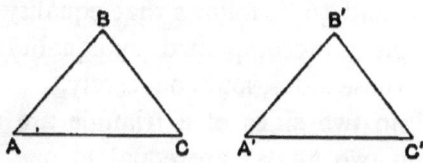

Proof.—Place A' on A, and A'C' along AC.

Because A'C'=AC, C' coincides with C; (27°)
and ∵ ∠A'=∠A, A'B' lies along AB;
and ∵ ∠C'=∠C, C'B' lies along CB;
∴ B' coincides with B, (24°, Cor. 3)
and the triangles are congruent. q.e.d.

60°. *Theorem.*—An external angle of a triangle is greater than an internal opposite angle.

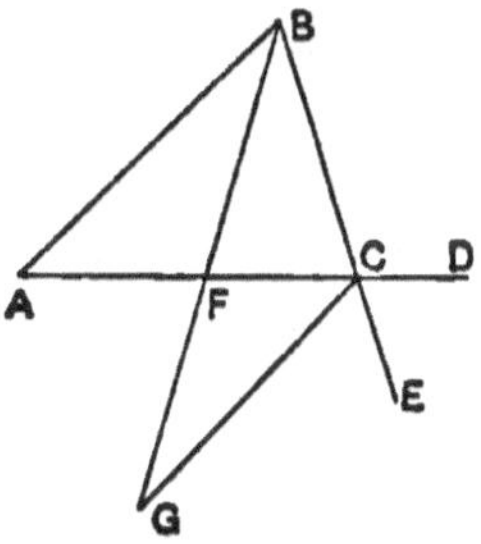

The external angle BCD is greater than the internal opposite angle ABC or BAC.

Proof.—Let BF be a median produced until FG=BF.

Then the △s ABF and CGF have
BF=FG, (construction)
AF=FC, (55°)
and ∠BFA=∠GFC. (40°)
∴ △ABF≡△CGF, (52°)
and ∠FCG=∠BAC. (52°, Cor.)
But ∠ACE is greater than ∠FCG.
∴ ∠ACE is > ∠BAC.
Similarly, ∠BCD is > ∠ABC,
and ∠BCD=∠ACE. (40°)

Therefore the ∠s BCD and ACE are each greater than each of the ∠s ABC and BAC. q.e.d.

61°. *Theorem.*—Only one perpendicular can be drawn to a line from a point not on the line.

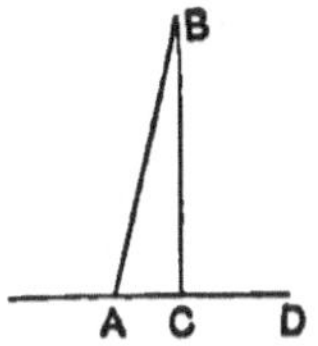

Proof.—Let B be the point and AD the line; and let BC be ⊥ to AD, and BA be any line other than BC.

Then ∠BCD is > ∠BAC, (60°)
∴ ∠BAC is not a ⊐,
and BA is not ⊥ to AD.
But BA is any line other than BC;
∴ BC is the only perpendicular from B to AD. q.e.d.

THREE OR MORE POINTS AND LINES.   31.

Cor. Combining this result with that of 42° we have—
Through a given point only one perpendicular can be drawn to a given line.

62°. *Theorem.*—Of any two unequal sides of a triangle and the opposite angles—
  1. The greater angle is opposite the longer side.
  2. The longer side is opposite the greater angle.

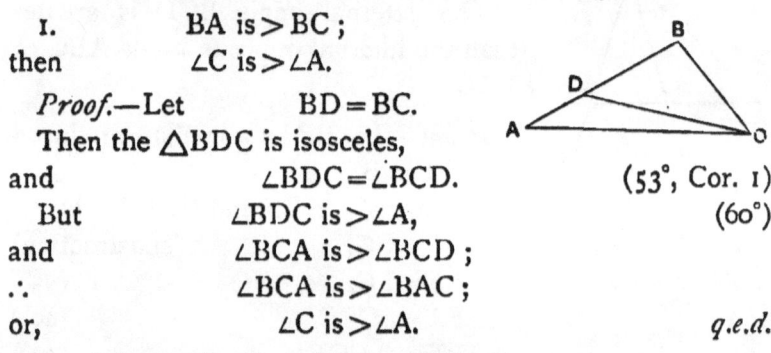

1.   BA is > BC ;
then   ∠C is > ∠A.

Proof.—Let   BD = BC.
Then the △BDC is isosceles,
and   ∠BDC = ∠BCD.   (53°, Cor. 1)
But   ∠BDC is > ∠A,   (60°)
and   ∠BCA is > ∠BCD ;
∴   ∠BCA is > ∠BAC ;
or,   ∠C is > ∠A.   *q.e.d.*

2. ∠C is > ∠A ; then AB is > BC.

Proof.—From the Rule of Identity (7°), since there is but one longer side and one greater angle, and since it is shown (1) that the greater angle is opposite the longer side, therefore the longer side is opposite the greater angle.   *q.e.d.*

Cor. 1. In any scalene triangle the sides being unequal to one another, the greatest angle is opposite the longest side, and the longest side is opposite the greatest angle.

Also, the shortest side is opposite the smallest angle, and conversely.

Hence if *A*, *B*, *C* denote the angles, and *a*, *b*, *c* the sides respectively opposite, the order of magnitude of A, B, C is the same as that of *a*, *b*, *c*.

63°. *Theorem.*—Of all the segments between a given point and a line not passing through the point—
  1. The perpendicular to the line is the shortest.
  2. Of any two segments the one which meets the line

32   SYNTHETIC GEOMETRY.

further from the perpendicular is the longer; and conversely, the longer meets the line further from the perpendicular than the shorter does.

3. Two, and only two segments can be equal, and they lie upon opposite sides of the perpendicular.

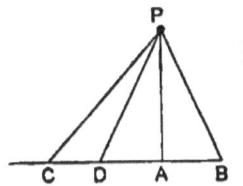

P is any point and BC a line not passing through it, and PA is ⊥ to BC.

1. PA which is ⊥ to BC is shorter than any segment PB which is not ⊥ to BC.

Proof.—	∠PAC=∠PAB=⌐	(hyp.)
But	∠PAC is > ∠PBC;	(60°)
∴	∠PAB is > ∠PBC,	
and ∴	PB is > PA.	(62°, 2) *q.e.d.*

2. AC is > AB, then also PC is > PB.

Proof.—Since AC is > AB, let D be the point in AC so that AD = AB.

Then A is the middle point of BD, and PA is the right bisector of BD.   (42°, Def.)

∴	PD = PB	(53°)
and	∠PDB = ∠PBD.	(53°, Cor 1)
But	∠PDB is > ∠PCB;	
∴	∠PBD is > ∠PCB,	
and	PC is > PB.	(62°, 2)

The converse follows from the Rule of Identity.   *q.e.d.*

3. *Proof.*—In 2 it is proved that PD = PB. Therefore two equal segments can be drawn from any point P to the line BC; and these lie upon opposite sides of PA.

No other segment can be drawn equal to PD or PB. For it must lie upon the same side of the perpendicular, PA, as one of them. If it lies further from the perpendicular than this one it is longer, (2), and if it lies nearer the perpendicular it is shorter. Therefore it must coincide with one of them and is not a third line.   *q.e.d*

THREE OR MORE POINTS AND LINES.

Def.—The length of the perpendicular segment between any point and a line is the *distance* of the point from the line.

64°. *Theorem.*—If two triangles have two angles in the one respectively equal to two angles in the other, and a side opposite an equal angle in each equal, the triangles are congruent.

If $\angle A' = \angle A$
$\angle C' = \angle C$
and $A'B' = AB$ } then the $\triangle$s $A'B'C'$ and ABC are congruent.

Proof.—Place A' on A, and A'B' along AB.

∵ $A'B' = AB$, B' coincides with B.   (27°)
Also, ∵ $\angle A' = \angle A$, A'C' lies along AC.   (34°)

Now if C' does not coincide with C, let it fall at some other point, D, on AC.

Then, ∵ $AB = A'B'$, $AD = A'C'$, and $\angle A = \angle A'$,
∴ $\triangle A'B'C' \equiv \triangle ABD$,   (52°)
and $\angle ADB = \angle C'$.   (52°, Cor.)
But $\angle C' = \angle C$,   (hyp.)
∴ $\angle ADB = \angle C$,

which is not true unless D coincides with C.

Therefore C' must fall at C, and the $\triangle$s ABC and A'B'C' are congruent.

The case in which D may be supposed to be a point on AC produced is not necessary. For we may then superimpose the $\triangle$ABC on the $\triangle$A'B'C'.

65°. *Theorem.*—If two triangles have two sides in the one respectively equal to two sides in the other, and an angle opposite an equal side in each equal, then—

1. If the equal angles be opposite the longer of the two sides in each, the triangles are congruent.
2. If the equal angles be opposite the shorter of the two

sides in each, the triangles are not necessarily congruent.

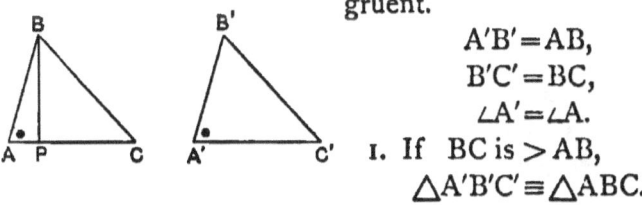

A′B′ = AB,
B′C′ = BC,
∠A′ = ∠A.

1. If BC is > AB,
△A′B′C′ ≡ △ABC.

Proof.—Since BC is > AB, therefore B′C′ is > A′B′. Place A′ on A and A′C′ along AC.

∵ ∠A′ = ∠A, and A′B′ = AB,
∴ B′ coincides with B. (34°, 27°)
Let BP be ⊥ AC;
then B′C′ cannot lie between BA and BP (63°, 2), but must lie on the same side as BC; and being equal to BC, the lines B′C′ and BC coincide (63°, 3), and hence
△A′B′C′ ≡ △ABC. *q.e.d.*

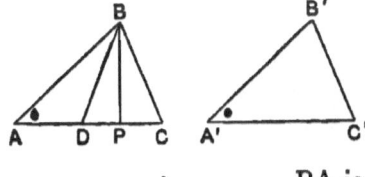

2. If BC is < AB, the △s A′B′C′ and ABC may or may not be congruent.

Proof.—Since AB is > BC,
∴ PA is > PC. (63°, 2)
Let PD = PC,
then BD = BC.

Now, let △A′B′C′ be superimposed on △ABC so that A′ coincides with A, B′ with B, and A′C′ lies along AC. Then, since we are not given the length of A′C′, B′C′ may coincide with BC, and the △s A′B′C′ and ABC be congruent; or B′C′ may coincide with BD, and the triangles A′B′C′ and ABC be not congruent. *q.e.d.*

Hence when two triangles have two sides in the one respectively equal to two sides in the other, and an angle opposite one of the equal sides in each equal, the triangles are not necessarily congruent unless some other relation exists between them.

The first part of the theorem gives one of the sufficient relations. Others are given in the following corollaries.

Cor. 1. If ∠C is a ⌐, BC and BD (2nd Fig.) coincide along BP, and the △s ABD and ABC become one and the same. Hence C′ must fall at C, and the △s A′B′C′ and ABC are congruent.

Cor. 2. The ∠BDA is supplementary to BDC and therefore to BCA. And ∵ ∠BDA is > ∠BPA, ∴ ∠BDA is greater than a right angle, and the ∠BCA is less than a right angle.

Hence if, in addition to the equalities of the theorem, the angles C and C′ are both equal to, or both greater or both less than a right angle, the triangles are congruent.

Def.—Angles which are both greater than, or both equal to, or both less than a right angle are said to be *of the same affection*.

66°. A triangle consists of six parts, three sides and three angles. When two triangles are congruent all the parts in the one are respectively equal to the corresponding parts in the other. But in order to establish the congruence of two triangles it is not *necessary* to establish independently the respective equality of all the parts; for, as has now been shown, if certain of the corresponding parts be equal the equality of the remaining parts and hence the congruence of the triangles follow as a consequence. Thus it is sufficient that two sides and the included angle in one triangle shall be respectively equal to two sides and the included angle in another. For, if we are given these parts, we are given consequentially all the parts of a triangle, since every triangle having two sides and the included angle equal respectively to those given is congruent with the given triangle.

Hence a triangle is *given* when two of its sides and the angle between them are given.

A triangle is *given* or *determined* by its elements being given according to the following table :—

1. Three sides, ... (58°)
2. Two sides and the included angle, ... (52°)
3. Two angles and the included side, ... (59°)
4. Two angles and an opposite side, ... (64°)
5. Two sides and the angle opposite the longer side, (65°)

When the three parts given are two sides and the angle opposite the shorter side, two triangles satisfy the conditions, whereof one has the angle opposite the longer side supplementary to the corresponding angle in the other.

This is known as the *ambiguous case* in the solution of triangles.

A study of the preceding table shows that a triangle is completely given when any three of its six parts are given, with two exceptions :—

(1) The three angles ;
(2) Two sides and the angle opposite the shorter of the two sides.

67°. *Theorem.*—If two triangles have two sides in the one respectively equal to two sides in the other, but the included angles and the third sides unequal, then

1. The one having the greater included angle has the greater third side.
2. Conversely, the one having the greater third side has the greater included angle.

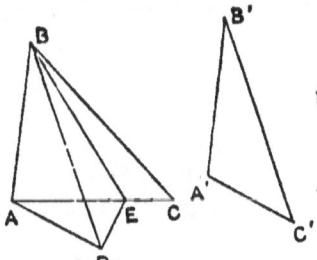

A'B' = AB and B'C' = BC, and
1. ∠ABC is > ∠A'B'C',
then   AC is > A'C'.

Proof.—Let A' be placed upon A and A'B' along AB.
Since A'B' = AB, B' falls on B.
Let C' fall at some point D.
Then ABD is A'B'C' in its new position.

Let BE bisect the ∠DBC and meet AC in E.
Join DE.
Then, in the △s DBE and CBE,
$$DB = BC, \qquad \text{(hyp.)}$$
$$\angle DBE = \angle CBE, \qquad \text{(constr.)}$$
and BE is common.
∴ $\triangle DBE \equiv \triangle CBE$,
and $DE = CE$.
But $AC = AE + EC = AE + ED$,
which is greater than AD.
∴ AC is > A′C′. *q.e.d.*

2. AC is > A′C′, then ∠ABC is greater than ∠A′B′C′.

Proof.—The proof of this follows from the Rule of Identity.

68°. *Theorem*.—1. Every point upon a bisector of an angle is equidistant from the arms of the angle.

2. Conversely, every point equidistant from the arms of an angle is on one of the bisectors of the angle.

1. OP and OQ are bisectors of the angle AOB, and PA, PB are perpendiculars from P upon the arms. Then
$$PA = PB.$$

Proof.—The △s POA and POB are congruent, since they have two angles and an opposite side equal in each (64°); ∴ $PA = PB.$

If Q be a point on the bisector OQ it is shown in a similar manner that the perpendiculars from Q upon the arms of the angle AOB are equal. *q.e.d.*

2. If PA is ⊥ to OA and PB is ⊥ to OB, and PA = PB, then PO is a bisector of the angle AOB.

Proof.—The △s POA and POB are congruent, since they have two sides and an angle opposite the longer equal in each (65°, 1) ; ∴ ∠POA = ∠POB,
and PO bisects the ∠AOB.

Similarly, if the perpendiculars from Q upon OA and OB are equal, QO bisects the ∠BOA′, or is the external bisector of the ∠AOB. *q.e.d.*

LOCUS.

69°. A *locus* is the figure traced by a variable point, which takes all possible positions subject to some constraining condition.

If the point is confined to the plane the locus is one or more lines, or some form of curve.

Illustration.—In the practical process of drawing a line or curve by a pencil, the point of the pencil becomes a variable (physical) point, and the line or curve traced is its locus.

In geometric applications the point, known as the *generating* point, moves according to some law.

The expression of this law in the Symbols of Algebra is known as the equation to the locus.

Cor. 1. The locus of a point in the plane, equidistant from the end-points of a given line-segment, is the right bisector of that segment.

This appears from 54°.

Cor. 2. The locus of a point in the plane, equidistant from two given lines, is the two bisectors of the angle formed by the lines.

This appears from 68°, converse.

EXERCISES.

1. How many lines at most are determined by 5 points? by 6 points? by 12 points?
2. How many points at most are determined by 6 lines? by 12 lines?
3. How many points are determined by 6 lines, three of which pass through a common point?

4. How many angles altogether are about a triangle? How many at most of these angles are different in magnitude? What is the least number of angles of different magnitudes about a triangle?
5. In Fig. of 53°, if Q be any point on PC, $\triangle PAQ \equiv \triangle PBQ$.
6. In Fig. of 53°, if the $\triangle PCB$ be revolved about PC as an axis, it will become coincident with $\triangle PCA$.
7. The medians to the sides of an isosceles triangle are equal to one another.
8. Prove 58° from the axiom. "a straight line is the shortest distance between two given points."
9. Show from 60° that a triangle cannot have two of its angles right angles.
10. If a triangle has a right angle, the side opposite that angle is greater than either of the other sides.
11. What is the locus of a point equidistant from two sides of a triangle?
12. Find the locus of a point which is twice as far from one of two given lines as from the other.
13. Find the locus of a point equidistant from a given line and a given point.

SECTION IV.

PARALLELS, ETC.

70°. *Def.*—Two lines, in the same plane, which do not intersect at any finite point are *parallel*.

Next to perpendicularity, parallelism is the most important directional relation. It is denoted by the symbol ||, which is to be read "parallel to" or "is parallel to" as occasion may require.

The idea of parallelism is identical with that of *sameness* of direction. Two line-segments may differ in length or in direction or in both.

If, irrespective of direction, they have the same length, they are equal; if, irrespective of length, they have the same direction, they are parallel; and if both length and direction are the same they are equal and parallel. Now when two segments are *equal* one may be made to coincide with the other by superposition without change of length, whether change of direction is required or not. So when they are parallel one may be made to coincide with the other without change of direction, whether change of length is required or not.

Axiom.—Through a given point only one line can be drawn parallel to a given line.

This axiom may be derived directly from 24°.

71°. *Theorem.*—Two lines which are perpendicular to the same line are parallel.

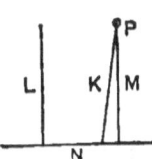

L and M are both ⊥ to N, then L is ∥ to M.

Proof.—If L and M meet at any point, two perpendiculars are drawn from that point to the line N.

But this is impossible (61°).

Therefore L and M do not meet, or they are parallel.

Cor. All lines perpendicular to the same line are parallel to one another.

72°. *Theorem.*—Two lines which are parallel are perpendicular to the same line, or they have a common perpendicular. (Converse of 71°.)

L is ∥ to M, and L is ⊥ to N; then M is ⊥ to N.

Proof.—If M is not ⊥ to N, through any point P in M, let K be ⊥ to N.

PARALLELS, ETC.

Then K is ∥ to L. (71°)
But M is ∥ to L. (hyp.)
Therefore K and M are both ∥ to L, which is impossible unless K and M coincide. (70°, Ax.)
Therefore L and M are both ⊥ to N, or N is a common perpendicular.

73°. *Def.*—A line which crosses two or more lines of any system of lines is a *transversal*. Thus EF is a transversal to the lines AB and CD.

In general, the angles formed by a transversal to any two lines are distinguished as follows—

a and e, c and g, b and f, d and h are pairs of corresponding angles.

c and f, e and d are pairs of alternate angles.

c and e, d and f are pairs of interadjacent angles.

74°. When a transversal crosses parallel lines—
1. The alternate angles are equal in pairs.
2. The corresponding angles are equal in pairs.
3. The sum of a pair of interadjacent angles is a straight angle.

AB is ∥ to CD and EF is a transversal.

1. ∠AEF = ∠EFD.

Proof.—Through O, the middle point of EF, draw PQ a common ⊥ to AB and CD. (72°)
Then △OPE ≡ △OQF ; (64°)
∴ ∠AEF = ∠EFD. *q.e.d.*

Similarly the remaining alternate angles are equal.

2. ∠AEG = ∠CFE, etc.

Proof.— ∠AEG = supplement of ∠AEF, (40°, Def. 1)
and ∠CFE = supplement of ∠EFD.

But $\angle AEF = \angle EFD$; (74°, 1)
∴ $\angle AEG = \angle CFE$. q.e.d.

Similarly the other corresponding angles are equal in pairs.

3. $\angle AEF + \angle CFE = \perp$.

Proof.— $\angle AEF = \angle EFD$, (74°, 1)
and $\angle CFE + \angle EFD = \perp$; (38°)
∴ $\angle AEF + \angle CFE = \perp$. q.e.d.

Cor. It is seen from the theorem that the equality of a pair of alternate angles determines the equality in pairs of corresponding angles, and also determines that the sum of a pair of interadjacent angles shall be a straight angle. So that the truth of any one of the statements 1, 2, 3 determines the truth of the other two, and hence if any one of the statements be proved the others are indirectly proved also.

75°. *Theorem.*—If a transversal to two lines makes a pair of alternate angles equal, the two lines are parallel. (Converse of 74° in part.)

If $\angle AEF = \angle EFD$, AB and CD are parallel.

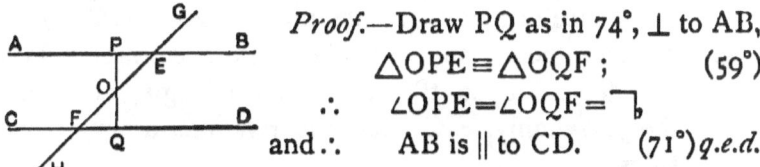

Proof.—Draw PQ as in 74°, $\perp$ to AB,
$\triangle OPE \equiv \triangle OQF$; (59°)
∴ $\angle OPE = \angle OQF = \perp$,
and ∴ AB is ∥ to CD. (71°) q.e.d.

Cor. It follows from 74° Cor. that if a pair of corresponding angles are equal to one another, or if the sum of a pair of interadjacent angles is a straight angle, the two lines are parallel.

76°. *Theorem.*—The sum of the internal angles of a triangle is a straight angle.

ABC is a $\triangle$;
the $\angle A + \angle B + \angle C = \perp$.

Proof.—Let CE be ∥ to AB, and D be any point on AC produced.

PARALLELS, ETC.

Then BC is a transversal to the parallels AB and CE ;
∴ $\angle ABC = \angle BCE$. (74°, 1)
Also, AC is a transversal to the same parallels ;
∴ $\angle BAC = \angle ECD$. (74°, 2)
∴ $\angle ABC + \angle BAC = \angle BCD$
= supplement of $\angle BCA$.
∴ $\angle A + \angle B + \angle C = \perp$. *q.e.d.*

Cor. An external angle of any triangle is equal to the sum of the opposite internal angles. (49°, 3)
For $\angle BCD = \angle A + \angle B$.

77°. From the property that the sum of the three angles of any triangle is a straight angle, and therefore constant, we deduce the following—

1. When two angles of a triangle are given the third is given also; so that the giving of the third furnishes no new information.
2. As two parts of a triangle are not sufficient to determine it, a triangle is not determined by its three angles, and hence one side, at least, must be given (66°, 1).
3. The magnitude of any particular angle of a triangle does not depend upon the *size* of the triangle, but upon the *form* only, *i.e.*, upon the relations amongst the sides.
4. Two triangles may have their angles respectively equal and not be congruent. But such triangles have the same form and are said to be *similar*.
5. A triangle can have but one obtuse angle; it is then called an *obtuse-angled* triangle.
    A triangle can have but one right angle, when it is called a *right-angled* triangle.
    All other triangles are called *acute-angled* triangles, and have three acute angles.
6. The acute angles in a right-angled triangle are complementary to one another.

78°. *Theorem.*—If a line cuts a given line it cuts every parallel to the given line.

Let L cut M, and let N be any parallel to M. Then L cuts N.

Proof.—If L does not cut N it is ∥ to N. But M is ∥ to N. Therefore through the same point P two lines L and M pass which are both ∥ to N.

But this is impossible; (70°, Ax.)

∴ L cuts N.

And N is any line ∥ to M.

∴ L cuts every line ∥ to M. *q.e.d.*

79°. *Theorem.*—If a transversal to two lines makes the sum of a pair of interadjacent angles less than a straight angle, the two lines meet upon that side of the transversal upon which these interadjacent angles lie.

GH is a transversal to AB and CD,

and ∠BEF + ∠EFD < ⊥.

Then AB and CD meet towards B and D.

Proof.—Let LK pass through E making ∠KEF = ∠EFC.

Then LK is ∥ to CD.

But AB cuts LK in E,

∴ it cuts CD. (78°)

Again, ∵ EB lies between the parallels, and AE does not, the point where AB meets CD must be on the side BD of the transversal. *q.e.d.*

Cor. Two lines, which are respectively perpendicular to two intersecting lines, intersect at some finite point.

80°. *Def.*—1. A closed figure having four lines as sides is in general called a *quadrangle* or *quadrilateral.*

Thus ABCD is a *quadrangle.*

2. The line-segments AC and BD which join opposite vertices are the *diagonals* of the quadrangle.

3. The quadrangle formed when two parallel lines intersect two other parallel lines is a *parallelogram*, and is usually denoted by the symbol ▱.

81°. *Theorem.*—In any parallelogram—
1. The opposite sides are equal to one another.
2. The opposite internal angles are equal to one another.
3. The diagonals bisect one another.

AB is ∥ to CD, and AC is ∥ to BD, and AD and BC are diagonals.

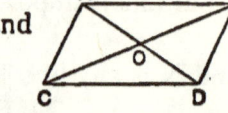

1. Then AB=CD and AC=BD.

Proof.—∵ AD is a transversal to the parallels AB and CD,
∴ ∠CDA=∠DAB.   (74°, 1)
and ∵ AD is a transversal to the parallels AC and BD,
∴ ∠CAD=∠ADB.   (74°, 1)
Hence, △CAD≡△BDA.   (59°)
∴ AB=CD and AC=BD.   *q.e.d.*

2. ∠CAB=∠BDC and ∠ACD=∠DBA.

Proof.—It is shown in 1 that ∠CAD=∠ADB and ∠BAD=∠ADC;
∴ by adding equals to equals,
∠CAB=∠CDB.
Similarly, ∠ACD=∠ABD.   *q.e.d.*

3. AO=OD and BO=OC.

Proof.—The △AOC≡△DOB;   (59°)
∴ AO=OD and BO=OC.   *q.e.d.*

82°. *Def.* 1.—A parallelogram which has two adjacent sides equal is a *rhombus*.

Cor. 1. Since AB=BC (hyp.)
=DC (81°, 1)=AD.

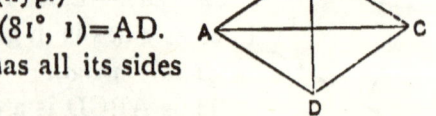

Therefore a rhombus has all its sides equal to one another.

Cor. 2. Since AC is the right bisector of BD,   (54°)
and BD the right bisector of AC,

Therefore the diagonals of a rhombus bisect one another at right angles.

Def. 2.—A parallelogram which has one right angle is a *rectangle*, and is denoted by the symbol ▭.

Cor. 3. Since the opposite angle is a ⌐, (81°, 2)
and the adjacent angle is a ⌐, (74°, 3)

Therefore a *rectangle* has all its angles right angles.

Cor. 4. The diagonals of a rectangle are equal to one another.

Def. 3.—A rectangle with two adjacent sides equal is a *square*, denoted by the symbol □.

Cor. 5. Since the square is a particular form of the rhombus and a particular form of the rectangle,

Therefore all the sides of a square are equal to one another; all the angles of a square are right angles; and the diagonals of a square are equal, and bisect each other at right angles.

84°. *Theorem.*—If three parallel lines intercept equal segments upon any one transversal they do so upon every transversal.

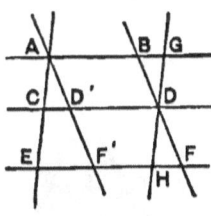

AE is a transversal to the three parallels AB, CD, and EF, so that $AC = CE$, and BF is any other transversal. Then $BD = DF$.

Proof.—Let GDH passing through D be ∥ to AE.

Then AGDC and CDHE are ▱s.		(80°, 3)
∴	$GD = AC = CE = DH$.	(81°, 1)
Also,	$\angle GBD = \angle DFH$, ∵ AG is ∥ to EF,	(74°, 1)
and	$\angle BDG = \angle FDH$;	(40°)
∴	$\triangle BDG \equiv \triangle FDH$,	(64°)
and	$BD = DF$.	q.e.d.

Def.—The figure ABFE is a *trapezoid*.
Therefore a trapezoid is a quadrangle having only two

PARALLELS, ETC.   47

sides parallel. The parallel sides are the major and minor *bases* of the figure.

Cor. 1. Since $2CD = AG + EH,$
$= AB + BG + EF - HF$
and $BG = HF;$
∴ $CD = \tfrac{1}{2}(AB + EF).$

Or, the line-segment joining the middle points of the non-parallel sides of a trapezoid is equal to one-half the sum of the parallel sides.

Cor. 2. When the transversals meet upon one of the extreme parallels, the figure AEF' becomes a △ and CD' becomes a line passing through the middle points of the sides AE and AF', and parallel to the base EF'.

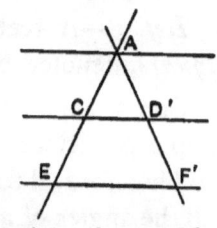

Therefore, 1, the line through the middle point of one side of a triangle, parallel to a second side, bisects the third side.

And, 2, the line through the middle points of two sides of a triangle is parallel to the third side.

85°. *Theorem.*—The three medians of a triangle pass through a common point.

CF and AD are medians intersecting in O.
Then BO is the median to AC.

Proof.—Let BO cut AC in E, and let AG ∥ to FC meet BO in G. Join CG.

Then, BAG is a △ and FO passes through the middle of AB and is ∥ to AG,

∴ O is the middle of BG.   (84°, Cor 2)

Again, DO passes through the middle points of two sides of the △CBG,

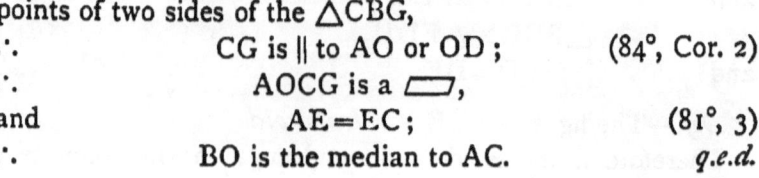

∴   CG is ∥ to AO or OD;   (84°, Cor. 2)
∴   AOCG is a ▱,
and   AE = EC;   (81°, 3)
∴   BO is the median to AC.   *q.e.d.*

Def.—When three or more lines meet in a point they are said to be *concurrent.*

Therefore the three medians of a triangle are concurrent.

Def. 2.—The point of concurrence, O, of the medians of a triangle is the *centroid* of the triangle.

Cor. Since O is the middle point of BG, and E is the middle point of OG, (81°, 3)
∴
$$OE = \tfrac{1}{2}OB,$$
$$= \tfrac{1}{3}EB.$$

Therefore the centroid of a triangle divides each median at two-thirds of its length from its vertex.

86°. *Theorem.*—The three right bisectors of the sides of a triangle are concurrent.

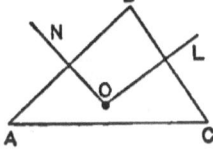

Proof.—Let L and N be the right bisectors of BC and AB respectively.

Then L and N meet in some point O.
(79°, Cor.)

Since L is the right bisector of BC, and N of AB, O is equidistant from B and C, and is also equidistant from A and B. (53°)

Therefore O is equidistant from A and C, and is on the right bisector of AC. (54°)

Therefore the three right bisectors meet at O. q.e.d.

Cor. Since two lines L and N can meet in only one point (24°, Cor. 3), O is the only point in the plane equidistant from A, B, and C.

Therefore only one finite point exists in the plane equidistant from three given points in the plane.

Def.—The point O, for reasons given hereafter, is called the *circumcentre* of the triangle ABC.

87°. *Def.*—The line through a vertex of a triangle perpendicular to the opposite side is *the perpendicular* to that side, and the part of that line intercepted within the triangle is the *altitude* to that side.

Where no reference to length is made the word *altitude* is often employed to denote the indefinite line forming the perpendicular.

Hence a triangle has three altitudes, one to each side.

88°. *Theorem.*—The three altitudes of a triangle are concurrent.

Proof.—Let ABC be a triangle. Complete the ▱s, ACBF, ABDC, and ABCE.
Then ∵ FB is ∥ to AC,
and BD is ∥ to AC,
FBD is one line, (70°, Ax.)
and FB = BD. (81°, 1)
Similarly, DCE is one line and DC = CE,
and EAF is one line and EA = AF.

Now, ∵ AC is ∥ to FD, the altitude to AC is ⊥ to FD and passes through B the middle point of FD. (72°)

Therefore the altitude to AC is the right bisector of FD, and similarly the altitudes to AB and BC are the right bisectors of DE and EF respectively.

But the right bisectors of the sides of the △DEF are concurrent (86°), therefore the altitudes of the △ABC are concurrent. *q.e.d.*

Def.—The point of concurrence of the altitudes of a triangle is the *orthocentre* of the triangle.

Cor. 1. If a triangle is acute-angled (77°, 5), the circumcentre and orthocentre both lie within the triangle.

2. If a triangle is obtuse-angled, the circumcentre and orthocentre both lie without the triangle.

3. If a triangle is right-angled, the circumcentre is at the middle point of the side opposite the right angle, and the orthocentre is the right-angled vertex.

Def.—The side of a right-angled triangle opposite the right angle is called the *hypothenuse*.

89°. The definition of 80° admits of three different figures, viz.:—

1. The *normal* quadrangle (1) in which each of the internal angles is less than a straight angle. When not

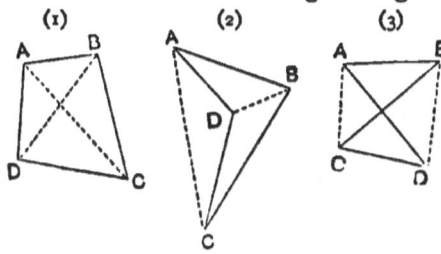

otherwise qualified the term quadrangle will mean this figure.

2. The quadrangle (2) in which one of the internal angles, as at D, is greater than a straight angle. Such an angle in a closed figure is called a *re-entrant* angle. We will call this an *inverted* quadrangle.

3. The quadrangle (3) in which two of the sides cross one another. This will be called a *crossed* quadrangle.

In each figure AC and BD are the diagonals; so that both diagonals are within in the normal quadrangle, one is within and one without in the inverted quadrangle, and both are without in the crossed quadrangle.

The general properties of the quadrangle are common to all three forms, these forms being only variations of a more general figure to be described hereafter.

90°. *Theorem.*—The sum of the internal angles of a quadrangle is four right angles, or a circumangle.

Proof.—The angles of the two △s ABD and CBD make up the internal angles of the quadrangle.

But these are $\perp + \perp$; (76°) therefore the internal angles of the quadrangle are together equal to four right angles. *q.e.d.*

Cor. This theorem applies to the inverted quadrangle as is readily seen.

PARALLELS, ETC.

91°. Theorem.—If two lines be respectively perpendicular to two other lines, the angle between the first two is equal or supplementary to the angle between the last two.

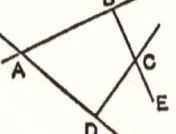

     BC is ⊥ to AB
and   CD is ⊥ to AD.

Then ∠(BC.CD) is equal or supplementary to ∠(AB.AD).

Proof.—ABCD is a quadrangle, and the ∠s at B and D are right angles; (hyp.)
∴   ∠BAD + ∠BCD = ⊥,   (90°)
or   ∠BCD is supplementary to ∠BAD.
But   ∠BCD is supplementary to ∠ECD;
and the ∠(BC.CD) is either the angle BCD or DCE. (39°)
∴ ∠(BC.CD) is = or supplementary to ∠BAD. *q.e.d.*

EXERCISES.

1. ABC is a △, and A′, B′, C′ are the vertices of equilateral △s described outwards upon the sides BC, CA, and AB respectively. Then AA′ = BB′ = CC′. (Use 52°.)
2. Is Ex. 1 true when the equilateral △s are described "inwardly" or upon the other sides of their bases?
3. Two lines which are parallel to the same line are parallel to one another.
4. L′ and M′ are two lines respectively parallel to L and M. The ∠(L′.M′) = ∠(L.M).
5. On a given line only two points can be equidistant from a given point. How are they situated with respect to the perpendicular from the given point?
6. Any side of a △ is greater than the difference between the other two sides.
7. The sum of the segments from any point within a △ to the three vertices is less than the perimeter of the △.
8. ABC is a △ and P is a point within on the bisector of ∠A. Then the difference between PB and PC is less than that between AB and AC, unless the △ is isosceles.

9. Is Ex. 8 true when the point P is without the △, but on the same bisector?
10. Examine Ex. 8 when P is on the external bisector of A, and modify the wording of the exercise accordingly.
11. CE and CF are bisectors of the angle between AB and CD, and EF is parallel to AB. Show that EF is bisected by CD.
12. If the middle points of the sides of a △ be joined two and two, the △ is divided into four congruent △s.
13. From any point in a side of an equilateral △ lines are drawn parallel to the other sides. The perimeter of the ▭ so formed is equal to twice a side of the △.
14. Examine Ex. 13 when the point is on a side produced.
15. The internal bisector of one angle of a △ and the external bisector of another angle meet at an angle which is equal to one-half the third angle of the △.
16. O is the orthocentre of the △ABC. Express the angles AOB, BOC, and COA in terms of the angles A, B, and C.
17. P is the circumcentre of the △ABC. Express the angles APB, BPC, and CPA in terms of the angles A, B, and C.
18. The joins of the middle points of the opposite sides of any quadrangle bisect one another.
19. The median to the hypothenuse of a right-angled triangle is equal to one-half the hypothenuse.
20. If one diagonal of a ▭ be equal to a side of the figure, the other diagonal is greater than any side.
21. If any point other than the point of intersection of the diagonals be taken in a quadrangle, the sum of the line-segments joining it with the vertices is greater than the sum of the diagonals.
22. If two right-angled △s have the hypothenuse and an acute angle in the one respectively equal to the like parts in the other, the △s are congruent.

PARALLELS, ETC.

23. The bisectors of two adjacent angles of a ▱ are ⊥ to one another.
24. ABC is a △. The angle between the external bisector of B and the side AC is ½(C∼A).
25. The external bisectors of B and C meet in D. Then ∠BDC = ½(B+C).
26. A line L which coincides with the side AB of the △ABC rotates about B until it coincides with BC, without at any time crossing the triangle. Through what angle does it rotate?
27. The angle required in Ex. 26 is an external angle of the triangle. Show in this way that the sum of the three external angles of a triangle is a circumangle, and that the sum of the three internal angles is a straight angle.
28. What property of space is assumed in the proof of Ex. 27?
29. Prove 76° by assuming that AC rotates to AB by crossing the triangle in its rotation, and that AB rotates to CB, and finally CB rotates to CA in like manner.

SECTION V.

THE CIRCLE.

92°. *Def.* 1.—A *Circle* is the locus of a point which, moving in the plane, keeps at a constant distance from a fixed point in the plane.

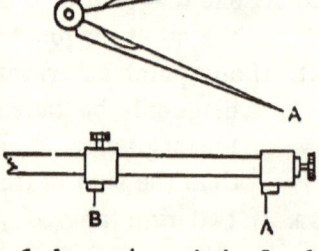

The compasses, whatever be their form, furnish us with two points, A and B, which, from the rigidity of the instrument, are supposed to preserve an unvarying distance from one another. Then, if one of the points A is fixed, while the other B moves over the paper or other plane surface, the moving point describes a physical circle. The limit of this physical circle, when the curved line has its thickness diminished endlessly, is the geometric circle.

Def. 2.—The fixed point is the *centre* of the circle, and the distance between the fixed and moveable points is the *radius* of the circle.

The curve itself, and especially where its length is under consideration, is commonly called the *circumference* of the circle.

The symbol employed for the circle is ⊙.

93°. From the definitions of 92° we deduce the following corollaries :—

1. All the radii of a ⊙ are equal to one another.

2. The ⊙ is a closed figure; so that to pass from a point within the figure to a point without it, or *vice versa*, it is necessary to cross the curve.

3. A point is within the ⊙, on the ⊙, or without the ⊙, according as its distance from the centre is less, equal to, or greater than the radius.

4. Two ⊙s which have equal radii are congruent; for, if the centres coincide, the figures coincide throughout and form virtually but one figure.

Def.—Circles which have their centres coincident are called *concentric* circles.

94°. *Theorem.*—A line can cut a circle in two points, and in two points only.

Proof. Since the ⊙ is a closed curve (93°, 2), a line which cuts it must lie partly within the ⊙ and partly without. And the generating point (69°) of the line must cross the ⊙ in passing from without to within, and again in passing from within to without.

∴ a line cuts a ⊙ at least twice if it cuts the ⊙ at all.

Again, since all radii of the same ⊙ are equal, if a line could cut a ⊙ three times, three equal segments could be drawn from a given point, the centre of the ⊙, to a given line. And this is impossible (63°, 3).

Therefore a line can cut a ⊙ only twice. q.e.d.

Cor. 1. Three points on the same circle cannot be in line; or, a circle cannot pass through three points which are in line.

95°. *Def.* 1.—A line which cuts a ⊙ is a *secant* or *secant-line*.

Def. 2.—The segment of a secant included within the ⊙ is a *chord*.

Thus the line L, or AB, is a secant, and the segment AB is a chord. (21°)

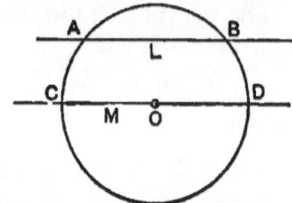

The term *chord* whenever involving the idea of length means the segment having its endpoints on the circle. But sometimes, when length is not involved, it is used to denote the whole secant of which it properly forms a part.

Def. 3.—A secant which passes through the centre is a *centre-line*, and its chord is a *diameter*.

Where length is not implied, the term *diameter* is sometimes used to denote the centre-line of which it properly forms a part.

Thus M is a centre-line and CD is a diameter.

96°. *Theorem.*—Through any three points not in line—
1. One circle can be made to pass.
2. Only one circle can be made to pass.

Proof.—Let A, B, C be three points not in line.

Join AB and BC, and let L and M be the right bisectors of AB and BC respectively.

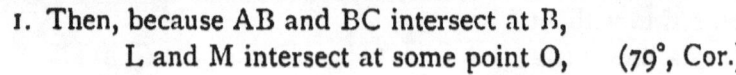

1. Then, because AB and BC intersect at B,
    L and M intersect at some point O,   (79°, Cor.)

and   O is equidistant from A, B, and C.   (86°)
∴ the ⊙ with centre at O, and radius equal to OA, passes through B and C.   *q.e.d.*

2. Any ⊙ through A, B, and C must have its centre equally distant from these three points.

But O is the only point in the plane equidistant from A, B, and C.   (86°, Cor.)

And we cannot have two separate ⊙s having the same centre and the same radius.   (93°, 4)

∴ only one circle can pass through A, B, and C. *q.e.d.*

Cor. 1. Circles which coincide in three points coincide altogether and form one circle.

Cor. 2. A point from which more than two equal segments can be drawn to a circle is the centre of that circle.

Cor. 3. Since L is a centre-line and is also the right bisector of AB,
∴ the right bisector of a chord is a centre line.

Cor. 4. The △AOB is isosceles, since OA=OB. Then, if D be the middle of AB, OD is a median to the base AB and is the right bisector of AB.   (55°, Cor. 2)
∴ a centre-line which bisects a chord is perpendicular to the chord.

Cor. 5. From Cor. 4 by the Rule of Identity,

A centre line which is perpendicular to a chord bisects the chord.

∴ the right bisector of a chord, the centre-line bisecting the chord, and the centre-line perpendicular to the chord are one and the same.

97°. From 92°, Def., a circle is given when the position of its centre and the length of its radius are given. And, from 96°, a circle is given when any three points on it are given.

It will be seen hereafter that a circle is determined by three points even when two of them become coincident, and in higher geometry it is shown that three points determine a circle, under certain circumstances, when all three of the points become coincident.

Def.—Any number of points so situated that a circle can pass through them are said to be *concyclic,* and a rectilinear figure (14°, Def.) having its vertices concyclic is said to be *inscribed* in the circle which passes through its vertices, and the circle is said to *circumscribe* the figure.

Hence the circle which passes through three given points is the *circumcircle* of the triangle having these points as vertices, and the centre of that circle is the *circumcentre* of the triangle, and its radius is the *circumradius* of the triangle.
(86°, Def.)

A like nomenclature applies to any rectilinear figure having its vertices concyclic.

98°. *Theorem.*—If two chords bisect one another they are both diameters.

If AP=PD and CP=PB, then P is the centre.

Proof.—Since P is the middle point of both AD and CB (hyp.), therefore the right bisectors of AD and CB both pass through P.

But these right bisectors also pass through the centre; (96°, Cor. 3)   ∴   P is the centre.   (24°, Cor. 3) *q.e.d.*

99°. *Theorem.*—Equal chords are equally distant from the centre ; and, conversely, chords equally distant from the centre are equal.

If AB=CD and OE and OF are the perpendiculars from the centre upon these chords, then OE=OF ; and conversely, if OE=OF, then AB=CD.

Proof.—Since OE and OF are centre lines ⊥ to AB and CD,
∴   AB and CD are bisected in E and F.   (96°, Cor. 5)
∴ in the △s OBE and ODF
$$OB=OD, \quad EB=FD,$$

58   SYNTHETIC GEOMETRY.

and they are right-angled opposite equal sides,
∴   △OBE ≡ △ODF,
and   OE = OF.

Conversely, by the Rule of Identity, if OE = OF, then AB = CD.   q.e.d.

100°. *Theorem.*—Two secants which make equal chords make equal angles with the centre-line through their point of intersection.

AB = CD, and PO is a centre-line through the point of intersection of AB and CD. Then
∠APO = ∠CPO.

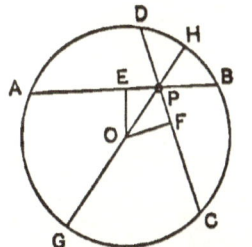

Proof.—Let OE and OF be ⊥ to AB and CD from the centre O.
Then   OE = OF,   (99°)
∴   △OPE ≡ △OPF,   (65°)
and   ∠APO = ∠CPO.   q.e.d.

Cor. 1. ∵ E and F are the middle points of AB and CD,   (96°, Cor. 5)
∴ PE = PF, PA = PC, and PB = PD.

Hence, secants which make equal chords make two pairs of equal line-segments between their point of intersection and the circle.

Cor. 2. From any point two equal line-segments can be drawn to a circle, and these make equal angles with the centre-line through the point.

101°. As all circles have the same form, two circles which have equal radii are equal and congruent (93°, 4), (51°). Hence *equal* and *congruent* are equivalent terms when applied to the circle.

Def. 1.—Any part of a circle is an *arc*.

The word *equal* when applied to arcs means congruence or capability of superposition. Equal arcs come from the same circle or from equal circles.

THE CIRCLE.

Def. 2.—A line which divides a figure into two parts such that when one part is revolved about the line it may be made to fall on and coincide with the other part is an *axis of symmetry* of the figure.

102°. *Theorem.*—A centre-line is an axis of symmetry of the circle.

Proof.—Let AB and CD be equal chords meeting at P, and let PHOG be a centre line.

Let the part of the figure which lies upon the F side of PG be revolved about PG until it comes to the plane on the E side of PG.

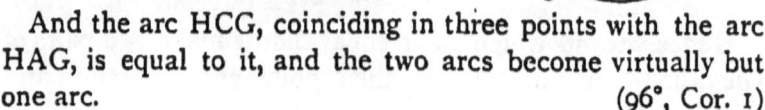

Then ∵ ∠GPA = ∠GPC, (100°)
∴ PC coincides with PA.
And ∵ PB = PD
and PA = PC, (100°, Cor. 1)
∴ D coincides with B,
and C coincides with A.

And the arc HCG, coinciding in three points with the arc HAG, is equal to it, and the two arcs become virtually but one arc. (96°, Cor. 1)

Therefore PG is an axis of symmetry of the ⊙, and divides it into two equal arcs.   *q.e.d.*

Def.—Each of the arcs into which a centre-line divides the circle is a *semicircle.*

Any chord, not a centre-line, divides the circle into unequal arcs, the greater of which is called the *major* arc, and the other the *minor* arc.

Cor. 1. By the superposition of the theorem we see that
  arc AB = arc CD, arc HB = arc HD, arc GA = arc GC,
∴   arc BDCA = arc DBAC.   (1st Fig.)

But the arcs BDCA and AB are the major and minor arcs

to the chord AB, and the arcs DBAC and CD are major and minor arcs to the chord CD.

Therefore equal chords determine equal arcs, major being equal to major and minor to minor.

Cor. 2. Equal arcs subtend equal angles at the centre.

103°. *Theorem.*—Parallel secants intercept equal arcs on a circle.

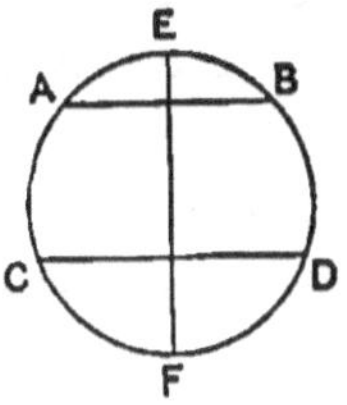

If   AB is ∥ to CD,
then   arc AC = arc DB.

Proof.—Let EF be the centre-line ⊥ to AB.
Then   EF is ⊥ to CD also.   (72°)
When EBDF is revolved about EF, B comes to coincidence with A, and D with C, and the arc BD with the arc AC,

∴   arc AC = arc DB.   *q.e.d.*

Cor. Since the chord AC = chord BD,

Therefore parallel chords have the chords joining their end-points equal.

EXERCISES.

1. Any plane closed figure is cut an even number of times by an indefinite line.
2. In the figure of Art. 96°, if A, B, and C shift their relative positions so as to tend to come into line, what becomes of the point O?
3. In the same figure, if ABC is a right angle where is the point O?
4. Given a circle or a part of a circle, show how to find its centre.
5. Three equal segments cannot be drawn to a circle from a point without it.
6. The vertices of a rectangle are concyclic.
7. If equal chords intersect, the segments of one between the

point of intersection and the circle are respectively equal to the corresponding segments of the other.
8. Two equal chords which have one end-point in common lie upon opposite sides of the centre.
9. If AB and CD be parallel chords, AD and BC, as also AC and BD, meet upon the right bisector of AB or CD.
10. Two secants which make equal angles with a centre-line make equal chords in the circle if they cut the circle.
(Converse of 100°.)
11. What is the axis of symmetry of (*a*) a square, (*b*) a rectangle, (*c*) an isosceles triangle, (*d*) an equilateral triangle? Give all the axes where there are more than one.
12. When a rectilinear figure has more than one axis of symmetry, what relation in direction do they hold to one another?
13. The vertices of an equilateral triangle trisect its circumcircle.
14. A centre-line perpendicular to a chord bisects the arcs determined by the chord.
15. Show how to divide a circle (*a*) into 6 equal parts, (*b*) into 8 equal parts.
16. If equal chords be in a circle, one pair of the connecters of their end-points are parallel chords.
(Converse of 103°, Cor.)

THE PRINCIPLE OF CONTINUITY.

104°. The principle of continuity is one of the most prolific in the whole range of Mathematics.

Illustrations of its meaning and application in Geometry will occur frequently in the sequel, but the following are given by way of introduction.

1. A magnitude is *continuous* throughout its extent.

Thus a line extends from any one point to another without

breaks or interruptions; or, a generating point in passing from one position to another must pass through *every* intermediate position.

2. In Art. 53° we have the theorem—Every point on the right bisector of a segment is equidistant from the end-points of the segment.

In this theorem the *limiting condition* in the hypothesis is that the point must be on the right bisector of the segment.

Now, if P be any point on the right bisector, and we move P along the right bisector, the limiting condition is not at any time violated during this motion, so that P remains *continuously* equidistant from the end-points of the segment during its motion.

We say then that the property expressed in the theorem is *continuous* while P moves along the right bisector.

3. In Art. 97° we have the theorem—The sum of the internal angles of a quadrangle is four right angles.

The limiting condition is that the figure shall be a quadrangle, and that it shall have internal angles.

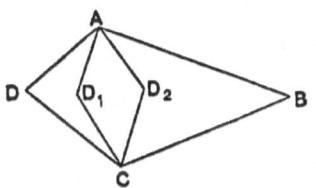

Now, let ABCD be a quadrangle. Then the condition is not violated if D moves to D_1 or D_2. But in the latter case the normal quadrangle ABCD becomes the inverted quadrangle $ABCD_2$, and the theorem remains true. Or, the theorem is continuously true while the vertex D moves anywhere in the plane, so long as the figure remains a quadrangle and retains four internal angles.

Future considerations in which a wider meaning is given to the word "angle" will show that the theorem is still true even when D, in its motion, crosses one of the sides AB or BC, and thus produces the crossed quadrangle.

The Principle of Continuity avoids the necessity of proving theorems for different *cases* brought about by variations in the disposition of the parts of a diagram, and it thus generalizes theorems or relieves them from dependence upon the particularities of a diagram. Thus the two figures of Art. 100° differ in that in the first figure the secants intersect without the circle, and in the second figure they intersect within, while the theorem applies with equal generality to both.

The Principle of Continuity may be stated as follows :—

When a figure, which involves or illustrates some geometric property, can undergo change, however small, in any of its parts or in their relations without violating the conditions upon which the property depends, then the property is *continuous* while the figure undergoes any amount of change of the same kind within the range of possibility.

105°. Let AB be a chord dividing the ⊙ into unequal arcs, and let P and Q be any points upon the major and minor arcs respectively. (102°, Def.)

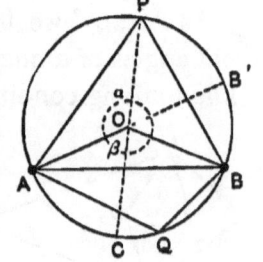

Let O be the centre.

1. The radii OA and OB form two angles at the centre, a *major* angle denoted by a and a *minor* angle denoted by β. These together make up a circumangle.

2. The chords PA, PB, and QA, QB form two angles at the circle, of which APB is the *minor* angle and AQB is the *major* angle.

3. The minor angle at the circle, APB, and the minor angle at the centre, β, stand upon the minor arc, AQB, as a base. Similarly the major angles stand upon the major arc as base.

4. Moreover the ∠APB is said to be *in the arc APB*, so that the minor angle at the circle is in the major arc, and the major angle at the circle is in the minor arc.

5. When B moves towards B' all the minor elements increase and all the major elements decrease, and when B comes to B' the minor elements become respectively equal to the major, and there is neither major nor minor.

When B, moving in the same direction, passes B', the elements change name, those which were formerly the minor becoming the major and *vice versa*.

106°. *Theorem*.—An angle at the circle is one-half the corresponding angle at the centre, major corresponding to major and minor to minor.

$\angle$AOB minor is 2$\angle$APB.

Proof.—Since $\triangle$APO is isosceles,

$\therefore$ $\angle$OAP = $\angle$OPA, (53°, Cor. 1)

and $\angle$OAP + $\angle$OPA = 2$\angle$OPA.

But $\angle$AOC = $\angle$OAP + $\angle$OPA,

(76°, Cor.)

$\therefore$ $\angle$AOC = 2$\angle$OPA.

Similarly $\angle$BOC = 2$\angle$OPB ;

$\therefore$ adding, $\angle$AOB minor = 2$\angle$APB. *q.e.d.*

The theorem is thus proved for the minor angles. But since the limiting conditions require only an angle at the circle and an angle at the centre, the theorem remains true while B moves along the circle. And when B passes B' the angle APB becomes the major angle at the circle, and the angle AOB minor becomes the major angle at the centre.

$\therefore$ the theorem is true for the major angles.

Cor. 1. The angle in a given arc is constant. (105°, 4)

Cor. 2. Since $\angle$APB = $\frac{1}{2}\angle$AOB minor,

and $\angle$AQB = $\frac{1}{2}\angle$AOB major,

and $\because$ $\angle$AOB minor + $\angle$AOB major = 4 right angles (37°)

$\therefore$ $\angle$APB + $\angle$AQB = a straight angle.

And APBQ is a concyclic quadrangle.

Hence a concyclic quadrangle has its opposite internal angles supplementary. (40°, Def. 1)

Cor. 3. D being on AQ produced,

$\angle$BQD is supplementary to $\angle$AQB.

But ∠APB is supplementary to ∠AQB,
∴ ∠APB = ∠BQD.

Hence, if one side of a concyclic quadrangle be produced, the external angle is equal to the opposite internal angle.

Cor. 4. Let B come to B'. (Fig. of 106°)
Then ∠AOB' is a straight angle,
∴ ∠APB' is a right angle.
But the arc APB' is a *semicircle*, (102°, Def.)
Therefore the angle in a semicircle is a right angle.

107°. *Theorem.*—A quadrangle which has its opposite angles supplementary has its vertices concyclic.

(Converse of 106°, Cor. 2)

ABCD is a quadrangle whereof the ∠ADC is supplementary to ∠ABC; then a circle can pass through A, B, C, and D.

Proof.—If possible let the ⊙ through A, B, and C cut AD in some point P.

Join P and C.

Then ∠APC is supplementary to ∠ABC, (106°, Cor. 2)
and ∠ADC is supplementary to ∠ABC, (hyp.)
∴ ∠APC = ∠ADC,
which is not true. (60°)
∴ the ⊙ cannot cut AD in any point other than D,
Hence A, B, C, and D are concyclic. *q.e.d.*

Cor. 1. The hypothenuse of a right-angled triangle is the diameter of its circumcircle. (88°, 3, Def.; 97°, Def.)

Cor. 2. When P moves along the ⊙ the △APC (last figure) has its base AC constant and its vertical angle APC constant.

Therefore the locus of the vertex of a triangle which has a constant base and a constant vertical angle is an arc of a circle passing through the end-points of the base.

This property is employed in the trammel which is used to describe an arc of a given circle.

E

66 SYNTHETIC GEOMETRY.

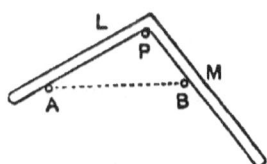

It consists of two rules (16°) L and M joined at a determined angle. When it is made to slide over two pins A and B, a pencil at P traces an arc passing through A and B.

108°. *Theorem.*—The angle between two intersecting secants is the sum of those angles in the circle which stand on the arcs intercepted between the secants, when the secants intersect within the circle, and is the difference of these angles when the secants intersect without the circle.

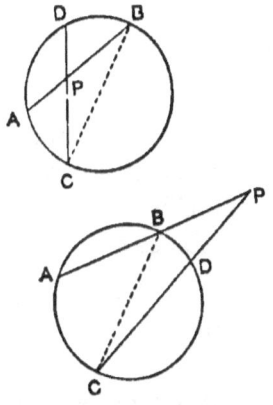

$\angle APC = \angle ABC + \angle BCD$, (1st Fig.)
$\angle APC = \angle ABC - \angle BCD$. (2nd Fig.)

Proof.—1. $\angle APC = \angle PBC + \angle PCB$, (60°)
∴ $\angle APC = \angle ABC + \angle BCD$.

2. $\angle ABC = \angle APC + \angle BCP$,
∴ $\angle APC = \angle ABC - \angle BCD$.

q.e.d.

EXERCISES.

1. If a six-sided rectilinear figure has its vertices concyclic, the three alternate internal angles are together equal to a circumangle.

2. In Fig. 105°, when B comes to Q, BQ vanishes; what is the direction of BQ just as it vanishes?

3. Two chords at right angles determine four arcs of which a pair of opposite ones are together equal to a semicircle.

4. A, B, C, D are the vertices of a square, and A, E, F of an equilateral triangle inscribed in the same circle. What is the angle between the lines BE and DF? between BF and ED?

SPECIAL SECANTS—TANGENT.

109°. Let P be a fixed point on the ⊙ S and Q a variable one.

The position of the secant L, cutting the circle in P and Q, depends upon the position of Q.

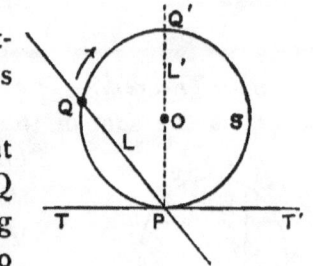

As Q moves along the ⊙ the secant rotates about P as pole. While Q makes one complete revolution along the ⊙ the secant L passes through two special positions. The first of these is when Q is farthest distant from P, as at Q', and the secant L becomes a *centre-line*. The second is when Q comes into coincidence with P, and the secant takes the position TT' and becomes a *tangent*.

Def. 1.—A *tangent* to a circle is a secant in its limiting position when its points of intersection with the circle become coincident.

That the tangent cannot cut or cross the ⊙ is evident. For if it cuts the ⊙ at P it must cut it again at some other point. And since P represents two points we would have the absurdity of a line cutting a circle in three points. (94°)

Def. 2.—The point where P and Q meet is called the *point of contact*. Being formed by the union of two points it represents both, and is therefore a *double* point.

From Defs. 1 and 2 we conclude—

1. A point of contact is a double point.
2. As a line can cut a ⊙ only twice it can touch a ⊙ only once.
3. A line which touches a ⊙ cannot cut it.
4. A ⊙ is determined by two points if one of them is a given point of contact on a given line; or, only one circle can pass through a given point and touch a given line at a given point. (Compare 97°.)

110°. *Theorem.*—A centre-line and a tangent to the same point on a circle are perpendicular to one another.

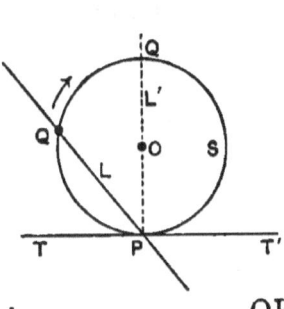

L′ is a centre-line and T a tangent, both to the point P. Then L′ is ⊥ to T.

Proof.—∵ T has only the one point P in common with the ⊙, every point of T except P lies without the ⊙. ∴ if O is the centre on the line L′, OP is the shortest segment from O to T. ∴ OP, or L′, is ⊥ to T.   (63°, 1) *q.e.d.*

Cor. 1. Tangents at the end-points of a diameter are parallel.

Cor. 2. The perpendicular to a tangent at the point of contact is a centre-line. (Converse of the theorem.)

Cor. 3. The perpendicular to a diameter at its end-point is a tangent.

111°. *Theorem.*—The angles between a tangent and a chord from the point of contact are respectively equal to the angles in the opposite arcs into which the chord divides the circle.

TP is a tangent and PQ a chord to the same point P, and A is any point on the ⊙. Then
∠QPT=∠QAP.

Proof.—Let PD be a diameter.
Then   ∠QAP=∠QDP,   (106°, Cor. 1)
and   ∠DQP is a ⌐.   (106°, Cor. 4)
Also   ∠DPQ is comp. of ∠QPT, (40°, Def. 3)
and   ∠DPQ is comp. of ∠QDP, (77°, 6)
∴   ∠QDP=∠QPT=∠QAP.
Similarly, the ∠QPT′=∠QBP.   *q.e.d.*

112°. *Theorem.*—Two circles can intersect in only two points.

Proof.—If they can intersect in three points, two circles can be made to pass through the same three points. But this is not true. (96°)

∴ two circles can intersect in only two points.

Cor. Two circles can touch in only one point. For a point of contact is equivalent to two points of intersection.

113°. *Theorem.*—The common centre-line of two intersecting circles is the right bisector of their common chord.

O and O' are the centres of S and S', and AB is their common chord. Then OO' is the right bisector of AB.

Proof.—Since $\quad$ AO = BO,
and $\quad\quad\quad\quad\quad$ AO' = BO',
$\quad$ ∴ O is on the right bisector of AB. $\quad\quad$ (54°)
Similarly $\quad$ O' is on the right bisector of AB,
$\quad$ ∴ OO' is the right bisector of AB.

Cor. 1. By the principle of continuity, OO' always bisects AB. Let the circles separate until A and B coincide. Then the circles touch and OO' passes through the point of contact.

Def.—Two circles which touch one another have *external contact* when each circle lies without the other, and *internal contact* when one circle lies within the other.

Cor. 2. Since OO' (Cor. 1) passes through the point of contact when the circles touch one another—

(*a*) When the distance between the centres of two circles is the sum of their radii, the circles have external contact.

(*b*) When the distance between the centres is the difference of the radii, the circles have internal contact.

(c) When the distance between the centres is greater than the sum of the radii, the circles exclude each other without contact.

(d) When the distance between the centres is less than the difference of the radii, the greater circle includes the smaller without contact.

(e) When the distance between the centres is less than the sum of the radii and greater than their difference, the circles intersect.

114°. *Theorem.*—From any point without a circle two tangents can be drawn to the circle.

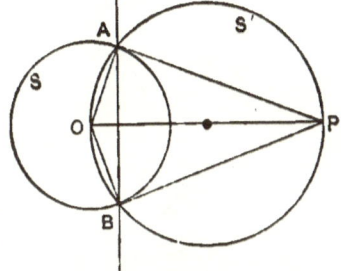

Proof.—Let S be the ⊙ and P the point. Upon the segment PO as diameter let the ⊙S' be described, cutting ⊙S in A and B. Then PA and PB are both tangents to S.

For ∠OAP is in a semicircle and is a ⌐, (106°, Cor. 4)
∴ AP is tangent to S. (110°, Cor. 3)
Similarly BP is tangent to S.

Cor. 1. Since PO is the right bisector of AB, (113°)
∴ PA = PB. (53°)

Hence calling the segment PA the *tangent from P to the circle,* when length is under consideration, we have—The two tangents from any point to a circle are equal to one another.

Def.—The line AB, which passes through the points of contact of tangents from P, is called the *chord of contact for the point P.*

115°. *Def.* 1.—The angle at which two circles intersect is the angle between their tangents at the point of intersection.

Def. 2. When two circles intersect at right angles they are said to cut each other *orthogonally.*

The same term is conveniently applied to the intersection of any two figures at right angles.

Cor. 1. If, in the Fig. to 114°, PA be made the radius of a circle and P its centre, the circle will cut the circle S orthogonally. For the tangents at A are respectively perpendicular to the radii.

Hence a circle S is cut orthogonally by any circle having its centre at a point without S and its radius the tangent from the point to the circle S.

116°. The following examples furnish theorems of some importance.

Ex. 1. Three tangents touch the circle S at the points A, B, and C, and intersect to form the $\triangle A'B'C'$. O being the centre of the circle,
$$\angle AOC = 2\angle A'OC'.$$

Proof.— $\quad AC' = BC',$
and $\qquad BA' = CA',$ (114°, Cor. 1)
∴ $\qquad \triangle AOC' \equiv \triangle BOC',$ and $\triangle BOA' \equiv \triangle COA'$
∴ $\qquad \angle BOC' = \angle AOC',$ and $\angle COA' = \angle BOA',$
∴ $\qquad \angle AOC = 2\angle A'OC'.$ $\qquad q.e.d.$
Similarly $\quad \angle AOB = 2\angle A'OB',$ and $\angle BOC = 2\angle B'OC'.$

If the tangents at A and C are fixed, and the tangent at B is variable, we have the following theorem :—

The segment of a variable tangent intercepted by two fixed tangents, all to the same circle, subtends a fixed angle at the centre.

Ex. 2. If four circles touch two and two externally, the points of contact are concyclic.

Let A, B, C, D be the centres of the circles, and P, Q, R, S be the points of contact.

Then AB passes through P, BC through Q, etc. (113°, Cor. 1)

Now, ABCD being a quadrangle,
$$\angle A + \angle B + \angle C + \angle D = 4 \text{ rt. } \angle\text{s}. \qquad (90°)$$
But the sum of all the internal angles of the four $\triangle$s APS, BQP, CRQ, and DSR is 8 rt. $\angle$s, and
$$\angle APS = \angle ASP, \angle BPQ = \angle BQP, \text{etc.},$$
$\therefore \quad \angle APS + \angle BPQ + \angle CRQ + \angle DRS = 2 \text{ rt. } \angle\text{s}.$
Now $\quad \angle SPQ = 2 \text{ rt. } \angle\text{s} - (\angle APS + \angle BPQ),$
$\quad\quad\quad \angle QRS = 2 \text{ rt. } \angle\text{s} - (\angle CRQ + \angle DRS),$
$\therefore \quad \angle SPQ + \angle QRS = 2 \text{ rt. } \angle\text{s}.$
and P, Q, R, S are concyclic (107°). *q.e.d.*

Ex. 3. If the common chord of two intersecting circles subtends equal angles at the two circles, the circles are equal.

AB is the common chord, C, C′ points upon the circles, and
$$\angle ACB = \angle AC'B.$$
Let O, O′ be the centres. Then $\angle AOB = \angle AO'B$. (106°) And the triangles OAB and O′AB being isosceles are congruent, $\therefore$ OA = O′A, and the circles are equal. (93°, 4)

Ex. 4. If O be the orthocentre of a $\triangle$ABC, the circumcircles to the $\triangle$s ABC, AOB, BOC, COA are all equal.

$\because$ AX and CZ are $\perp$ respectively to BC and AB,
$$\angle CBA = \text{sup. of } \angle XOZ$$
$$= \text{sup. of } \angle COA.$$
But D being any point on the arc AS$_2$C, $\angle$CDA is the sup. of $\angle$COA.
$\therefore \quad \angle CBA = \angle CDA,$
and the $\odot$s S and S$_2$ are equal by Ex. 3. In like manner it may be proved that the $\odot$S is equal to the $\odot$s S$_3$ and S$_1$.

Ex. 5. If any point O be joined to the vertices of a $\triangle$ABC, the circles having OA, OB, and OC as diameters intersect upon the triangle.

Proof.—Draw OX $\perp$ to BC and OY $\perp$ to AC.

∴ $\angle OXB = \rceil$, the ⊙ on OB as diameter passes through X.
(107°, Cor. 1)

Similarly the ⊙ on OC as diameter passes through X. Therefore the ⊙s on OB and OC intersect in X; and in like manner it is seen that the ⊙s on OC and OA intersect in Y, and those on OA and OB intersect in Z, the foot of the ⊥ from O to AB.

Ex. 6. The feet of the medians and the feet of the altitudes in any triangle are six concyclic points, and the circle bisects that part of each altitude lying between the orthocentre and the vertex.

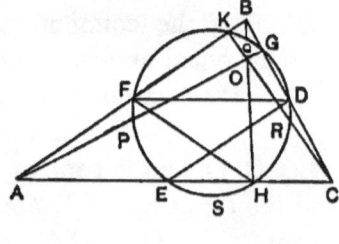

D, E, F are the feet of the medians, *i.e.*, the middle points of the sides of the △ABC. Let the circle through D, E, F cut the sides in G, H, K.

Now FD is ∥ to AC and ED is ∥ to AB, (84°, Cor. 2)
∴ $\angle FDE = \angle FAE$.
But $\angle FDE = \angle FHE$, (106°, Cor. 1)
∴ △AFH is isosceles, and AF = FH = FB;
∴ $\angle AHB = \rceil$, (106°, Cor. 4)
and H is the foot of the altitude from B.

Similarly, K and G are feet of the altitudes from C and A.

Again, $\angle KPH = \angle KFH = 2\angle KAH$. And A, K, O, H are concyclic (107°), and AO is a diameter of the circumcircle, therefore P is the middle point of AO.

Similarly, Q is the middle point of BO, and R of CO.

Def.—The circle S passing through the nine points D, E, F, G, H, K, and P, Q, R, is called the *nine-points circle* of the △ABC.

Cor. Since the nine-points circle of ABC is the circumcircle of △DEF, whereof the sides are respectively equal to half the sides of the △ABC, therefore the radius of the nine-points circle of any triangle is one-half that of its circumcircle.

Exercises.

1. In 105° when P passes B where is the ∠APB?
2. A, B, C, D are four points on a circle whereof CD is a diameter and E is a point on this diameter. If ∠AEB = 2∠ACB, E is the centre.
3. The sum of the alternate angles of any octagon in a circle is six right angles.
4. The sum of the alternate angles of any concyclic polygon of $2n$ sides is $2(n-1)$ right angles.
5. If the angle of a trammel is 60° what arc of a circle will it describe? what if its angle is $n°$?
6. Trisect a right angle and thence show how to draw a regular 12-sided polygon in a circle.
7. If r, r' be the radii of two circles, and d the distance between them, the circles touch when $d = r \pm r'$.
8. Give the conditions under which two circles have 4, 3, 2, or 1 common tangent.
9. Prove Ex. 2, 116°, by drawing common tangents to the circles at P, Q, R, and S.
10. A variable chord passes through a fixed point on a circle, to find the locus of the middle point of the chord.
11. A variable secant passes through a fixed point, to find the locus of the middle point of the chord determined by a fixed circle.
12. In Ex. 11, what is the locus of the middle point of the secant between the fixed point and the circle?
13. In a quadrangle circumscribed to a circle the sums of the opposite sides are equal in pairs; and if the vertices be joined to the centre the sums of the opposite angles at the centre are equal in pairs.
14. If a hexagon circumscribe a circle the sum of three alternate sides is equal to that of the remaining three.
15. If two circles are concentric, any chord of the outer which is tangent to the inner is bisected by the point

of contact; and the parts intercepted on any secant between the two circles are equal to one another.

16. If two circles touch one another, any line through the point of contact determines arcs which subtend equal angles in the two circles.
17. If any two lines be drawn through the point of contact of two touching circles, the lines determine arcs whose chords are parallel.
18. If two diameters of two touching circles are parallel, the transverse connectors of their end-points pass through the point of contact.
19. The shortest chord that can be drawn through a given point within a circle is perpendicular to the centre-line through that point.
20. Three circles touch each other externally at A, B, and C. The chords AB and AC of two of the circles meet the third circle in D and E. Prove that DE is a diameter of the third circle and parallel to the common centre-line of the other two.
21. A line which makes equal angles with one pair of opposite sides of a concyclic quadrangle makes equal angles with the other pair, and also with the diagonals.
22. Two circles touch one another in A and have a common tangent BC. Then $\angle BAC$ is a right angle.
23. OA and OB are perpendicular to one another, and AB is variable in position but of constant length. Find the locus of the middle point of AB.
24. Two equal circles touch one another and each touches one of a pair of perpendicular lines. What is the locus of the point of contact of the circles?
25. Two lines through the common points of two intersecting circles determine on the circles arcs whose chords are parallel.
26. Two circles intersect in A and B, and through B a secant cuts the circles in C and D. Show that $\angle CAD$ is constant, the direction of the secant being variable.

27. At any point in the circumcircle of a square one of the sides subtends an angle three times as great as that subtended by the opposite side.
28. The three medians of any triangle taken in both length and direction can form a triangle.

SECTION VI.

CONSTRUCTIVE GEOMETRY,

INVOLVING THE PRINCIPLES OF THE FIRST FIVE SECTIONS, ETC.

117°. Constructive Geometry applies to the determination of geometric elements which shall have specified relations to given elements.

Constructive Geometry is *Practical* when the determined elements are physical, and it is *Theoretic* when the elements are supposed to be taken at their limits, and to be geometric in character. (12°)

Practical Constructive Geometry, or simply "Practical Geometry," is largely used by mechanics, draughtsmen, surveyors, engineers, etc., and to assist them in their work numerous aids known as "Mathematical Instruments" have been devised.

A number of these will be referred to in the sequel.

In "Practical Geometry" the "Rule" (16°) furnishes the means of constructing a line, and the "Compasses" (92°) of constructing a circle.

In Theoretic Constructive Geometry we *assume* the ability to construct these two elements, and by means of these we are to determine the required elements.

118°. To test the "Rule."

Place the rule on a plane, as at R, and draw a line AB

along its edge. Turn the rule into the position R'. If the edge now coincides with the line the rule is true.

This test depends upon the property that two finite points A and B determine one line.
(24°, Cor. 2)

Def.—A construction proposed is in general called a proposition (2°) and in particular a *problem*.

A complete problem consists of (1) the statement of what is to be done, (2) the construction, and (3) the proof that the construction furnishes the elements sought.

119°. *Problem.*—To construct the right bisector of a given line segment.

Let AB be the given segment.

Construction.—With A and B as centres and with a radius AD greater than half of AB describe circles.

Since AB is < the sum of the radii and > their difference, the circles will meet in two points P and Q. (113°, Cor. 2, *e*)

The line PQ is the right bisector required.

Proof.—P and Q are each equidistant from A and B and
∴ they are on the right bisector of AB ; (54°)
∴ PQ is the right bisector of AB.

Cor. 1. The same construction determines C, the middle point of AB.

Cor. 2. If C be a given point on a line, and we take A and B on the line so that CA=CB, then the right bisector of the segment AB passes through C and is ⊥ to the given line.

∴ the construction gives the perpendicular to a given line at a given point in the line.

120°. *Problem.*—To draw a perpendicular to a given line from a point not on the line.

Let L be the given line and P be the point.

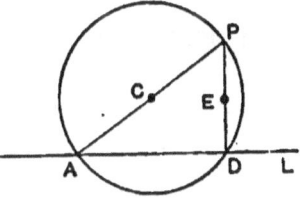

Constr.—Draw any line through P meeting L at some point A. Bisect AP in C (119°, Cor. 1), and with C as centre and CP as radius describe a circle.

If PA is not ⊥ to L, the ⊙ will cut L in two points A and D.

Then PD is the ⊥ required.

Proof.—PDA is the angle in a semicircle,

∴ ∠PDA is a ⌐. (106°, Cor. 4)

Cor. Let D be a given point in L. With any centre C and CD as radius describe a circle cutting L again in some point A. Draw the radius ACP, and join D and P. Then DP is ⊥ to L.

∴ the construction draws a ⊥ to L at a given point in L.

(Compare 119°, Cor. 2)

Cor. 2. Let L be a given line and C a given point.

To draw through C a line parallel to L.

With C as the centre of a circle, construct a figure as given. Bisect PD in E (119°, Cor. 1). Then CE is ∥ to L. For C and E are the middle points of two sides of a triangle of which L is the base. (84°, Cor. 2)

121°. *The Square.*—The square consists of two rules with their edges fixed permanently at right angles, or of a triangular plate of wood or metal having two of its edges at right angles.

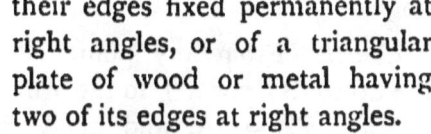

To test a square.

Draw a line AB and place the square as at S, so that one edge coincides with the line, and along the other edge draw the line CD.

Next place the square in the position S'. If the edges can

CONSTRUCTIVE GEOMETRY. 79

now be made to coincide with the two lines the square is true.

This test depends upon the fact that a right angle is one-half a straight angle.

The square is employed practically for drawing a line ⊥ to another line.

Cor. 1. The square is employed to draw a series of parallel lines, as in the figure.

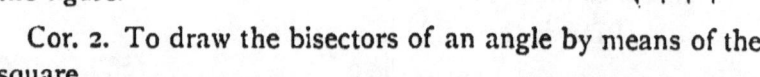

Cor. 2. To draw the bisectors of an angle by means of the square.

Let AOB be the given angle. Take OA=OB, and at A and B draw perpendiculars to OA and OB.

Since AOB is not a straight angle, these perpendiculars meet at some point C. (79°, Cor.)

Then OC is the internal bisector of ∠AOB. For the triangles AOC and BOC are evidently congruent.

∴   ∠AOC=∠BOC.

The line drawn through O ⊥ to OC is the external bisector.

122°. *Problem.*—Through a given point in a line to draw a line which shall make a given angle with that line.

Let P be the given point in the line L, and let X be the given angle.

Constr.— From any point B in the arm OB draw a ⊥ to the arm OA.   (120°)

Make PA'=OA, and at A' draw the ⊥ A'B' making A'B'=AB. PB' is the line required.

Proof.—The triangles OBA and PB'A' are evidently congruent, and ∴   ∠BOA=X=∠B'PA'.

Cor. Since PA' might have been taken to the left of P, the problem admits of two solutions. When the angle X is a right angle the two solutions become one.

123°. *The Protractor.*—This instrument has different forms depending upon the accuracy required of it. It usually consists of a semicircle of metal or ivory divided into degrees, etc. (41°). The point C is the centre. By placing the straight edge of the instrument in coincidence with a given line AB so that the centre falls at a given point C, we can set off any angle given in degrees, etc., along the arc as at D. Then the line CD passes through C and makes a given angle with AB.

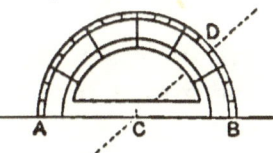

124°. *Problem.*—Given the sides of a triangle to construct it.

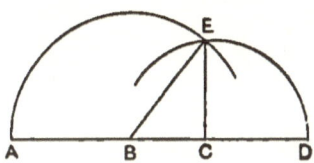

Constr.—Place the three sides of the triangle in line, as AB, BC, CD.

With centre C and radius CD describe a circle, and with centre B and radius BA describe a circle.

Let E be one point of intersection of these circles.

Then △BEC is the triangle required.

Proof.—BE = BA and CE = CD.

Since the circles intersect in another point E', a second triangle is formed. But the two triangles being congruent are virtually the same triangle.

Cor. 1. When AB = BC = CA the triangle is equilateral.

(53°, Def. 2)

In this case the circle AE passes through C and the circle DE through B, so that B and C become the centres and BC a common radius.

Cor. 2. When BC is equal to the sum or difference of AB and CD the circles touch (113°, Def.) and the triangle takes the limiting form and becomes a line.

When BC is greater than the sum or less than the difference of AB and CD the circles do not meet (113°, Def.) and no triangle is possible.

Therefore that three line-segments may form a triangle, each one must be less than the sum and greater than the difference of the other two.

125°. The solution of a problem is sometimes best effected by supposing the construction made, and then by reasoning backwards from the completed figure to some relation amongst the given parts by means of which we can make the construction.

This is analogous to the process employed for the solution of equations in Algebra, and a more detailed reference will be made to it at a future stage.

The next three problems furnish examples

126°. *Problem.*—To construct a triangle when two sides and the median to the third side are given.

Let a and b be two sides and n the median to the third side.

Suppose ACB is the required triangle having CD as the given median.

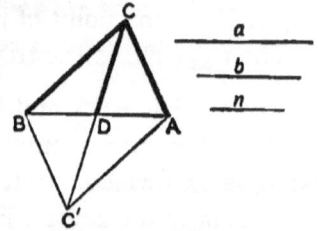

By completing the ▱ACBC' and joining DC', we have DC' equal to CD and in the same line, and BC'=AC (81°); and the triangle CC'B has CC'=$2n$, CB=a, and BC'=AC=b, and is constructed by 124°.

Thence the triangle ACB is readily constructed.

Cor. Since CC' is twice the given median, and since the possibility of the triangle ACB depends upon that of CC'B, therefore a median of a triangle is less than one-half the sum, and greater than one-half the difference of the conterminous sides. (124°, Cor. 2)

F

127°. *Problem.*—To trisect a given line-segment, *i.e.*, to divide it into three equal parts.

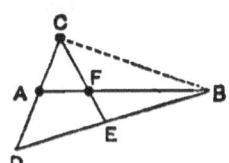

Constr.—Let AB be the segment.
Through A draw any line CD and make AC=AD. Bisect DB in E, and join CE, cutting AB in F.
Then AF is ⅓AB.

Proof.—CBD is a △ and CE and BA are two medians.
∴   AF=⅓AB.   (85°, Cor.)
Bisecting FB gives the other point of division.

128°. *Problem.*—To construct a △ when the three medians are given.

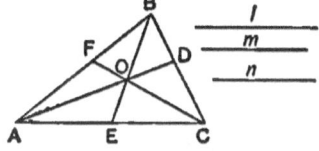

Let *l*, *m*, *n* be the given medians, and suppose ABC to be the required triangle. Then

AD=*l*, BE=*m*, and CF=*n*,
∴   AO=⅔*l*, OB=⅔*m*, and OF=⅓*n*,
∴ in the △AOB we have two sides and the median to the third side given. Thence △AOB is constructed by 126° and 127°.

Then producing FO until OC=2FO, C is the third vertex of the triangle required.

Ex. To describe a square whose sides shall pass through four given points.

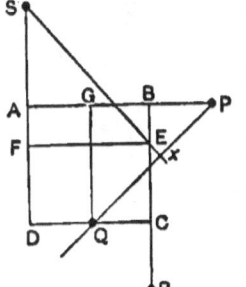

Let P, Q, R, S be the given points, and suppose ABCD to be the square required.
Join P and Q upon opposite sides of the square, and draw QG ∥ to BC. Draw SX ⊥ to PQ to meet BC in E, and draw EF ∥ to CD. Then △QPG ≡ △FSE,
and   SE=PQ.
Hence the construction :—
Join any two points PQ, and through a third point S draw

SX ⊥ to PQ. On SX take SE=PQ and join E with the fourth point R. ER is a side of the square in position and direction, and the points first joined, P and Q, are on opposite sides of the required square.

Thence the square is readily constructed.

Since SE may be measured in two directions along the line SX, two squares can have their sides passing through the same four points P, Q, R, S, and having P and Q on opposite sides.

Also, since P may be first connected with R or S, two squares can be constructed fulfilling the conditions and having P and Q on adjacent sides.

Therefore, four squares can be constructed to have their sides passing through the same four given points.

CIRCLES FULFILLING GIVEN CONDITIONS.

The problems occurring here are necessarily of an elementary character. The more complex problems require relations not yet developed.

129°. *Problem.*—To describe a circle to touch a given line at a given point.

P is a given point in the line L.

Constr.—Through P draw M ⊥ to L. A circle having any point C, on M, as centre and CP as radius touches L at P.

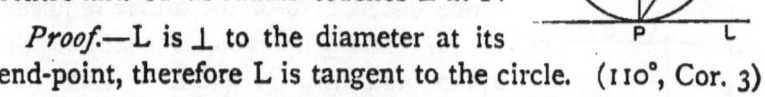

Proof.—L is ⊥ to the diameter at its end-point, therefore L is tangent to the circle. (110°, Cor. 3)

Def.—As C is *any* point on M, any number of circles may be drawn to touch L at the point P, and all their centres lie on M.

Such a problem is *indefinite* because the conditions are not sufficient to determine a *particular* circle. If the circle varies its radius while fulfilling the conditions of the problem, the centre moves along M; and M is called the *centre-locus* of the variable circle.

Hence the centre-locus of a circle which touches a fixed line at a fixed point is the perpendicular to the line at that point.

Cor. If the circle is to pass through a second given point Q the problem is definite and the circle is a particular one, since it then passes through three fixed points, viz., the double point P and the point Q. (109°, 4)

In this case ∠CQP=∠CPQ.

But ∠CPQ is given, since P, Q, and the line L are given.
∴ ∠CQP is given and C is a fixed point.

130°. *Problem.*—To describe a circle to touch two given non-parallel lines.

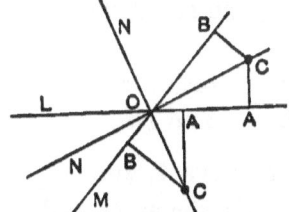

Let L and M be the lines intersecting at O.

Draw N, N, the bisectors of the angle between L and M. (121°, Cor. 2)

From C, any point on either bisector, draw CA ⊥ to L.

The circle with centre C and radius CA touches L, and if CB be drawn ⊥ to M, CB=CA. (68°)

Therefore the circle also touches M.

As C is any point on the bisectors the problem is indefinite, and the centre-locus of a circle which touches two intersecting lines is the two bisectors of the angle between the lines.

131°. *Problem.*—To describe a circle to touch three given lines which form a triangle.

L, M, N are the lines forming the triangle.

Constr.—Draw I_1, E_1, the internal and external bisectors of the angle A; and I_2, E_2, those of the angle B.

∵ ∠A+∠B is < ⊥, ∴ ∠BAO+∠ABO is < ⌐.

∴ I_1 and I_2 meet at some point O (79°) and are not ⊥ to one another and therefore E_1 and E_2 meet at some point O_3.
(79°, Cor.)

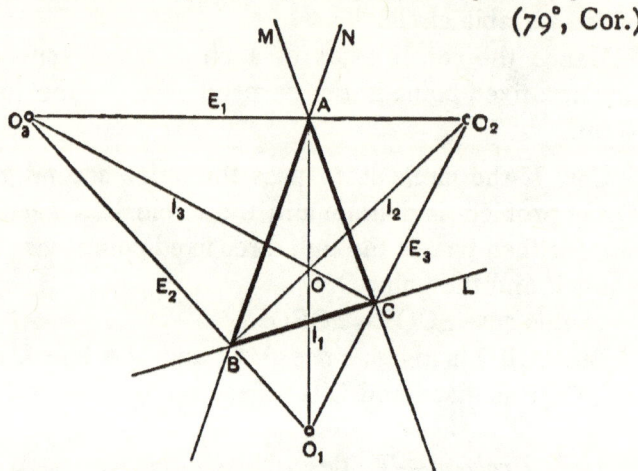

Also I_1 and E_2 meet at some point O_1, and similarly I_2 and E_1 meet at O_2.

The four points O, O_1, O_2, O_3 are the centres of four circles each of which touches the three lines L, M, and N.

Proof.—Circles which touch M and N have I_1 and E_1 as their centre-locus (130°), and circles which touch N and L have I_2 and E_2 as their centre-locus.

∴ Circles which touch L, M, and N must have their centres at the intersections of these loci.

But these intersections are O, O_1, O_2, and O_3,

∴ O, O_1, O_2, and O_3 are the centres of the circles required.

The radii are the perpendiculars from the centres upon any one of the lines L, M, or N.

Cor. 1. Let I_3 and E_3 be the bisectors of the ∠C. Then, since O is equidistant from L and M, I_3 passes through O. (68°)

∴ the three internal bisectors of the angles of a triangle are concurrent.

Cor. 2. Since O_3 is equidistant from L and M, I_3 passes through O_3. (68°)

∴ the external bisectors of two angles of a triangle and the internal bisector of the third angle are concurrent.

Def. 1.—When three or more points are in line they are said to be *collinear*.

Cor. 3. The line through any two centres passes through a vertex of the △ABC.

∴ any two centres are collinear with a vertex of the △.

The lines of collinearity are the six bisectors of the three angles A, B, and C.

Def. 2 —With respect to the △ABC, the circle touching the sides and having its centre at O is called the *inscribed* circle or simply the *in-circle* of the triangle.

The circles touching the lines and having centres at O_1, O_2, and O_3 are the *escribed* or *ex-circles* of the triangle.

REGULAR POLYGONS.

132°. *Def.* 1.—A closed rectilinear figure without re-entrant angles (89°, 2) is in general called a *polygon*.

They are named according to the number of their sides as follows :—

    3, triangle or trigon ;
    4, quadrangle, or tetragon, or quadrilateral ;
    5, pentagon ; 6, hexagon ; 7, heptagon ;
    8, octagon ; 10, decagon ; 12, dodecagon ; etc.

The most important polygons higher than the quadrangle are regular polygons.

Def. 2.—A *regular* polygon has its vertices concyclic, and all its sides equal to one another.

The centre of the circumcircle is the centre of the polygon.

133°. *Theorem.*—If n denotes the number of sides of a

regular polygon, the magnitude of an internal angle is $\left(2-\frac{4}{n}\right)$ right angles.

Proof.—Let AB, BC be two consecutive sides of the polygon and O its centre.

Then the triangles AOB, BOC are isosceles and congruent.

∴   ∠OAB=∠OBA=∠OBC=etc.,
∴   ∠OAB+∠OBA=∠ABC.
But   ∠OAB+∠OBA=⊥−∠AOB,
and (132°, Def. 2)   ∠AOB=$\frac{4 \text{ right angles}}{n}$,
∴   ∠ABC=$\left(2-\frac{4}{n}\right)$ right angles,
or   =$\left(2-\frac{4}{n}\right)$90°.   q.e.d.

Cor. The internal angles of the regular polygons expressed in right angles and in degrees are found, by putting proper values for n, to be as follows:—

Equilateral triangle,	$\frac{2}{3}$	60°	Octagon,	$\frac{3}{2}$	135°
Square,	1	90°	Decagon,	$\frac{8}{5}$	144°
Pentagon,	$\frac{6}{5}$	108°	Dodecagon,	$\frac{5}{3}$	150°
Hexagon,	$\frac{4}{3}$	120°			

134°. *Problem.*—On a given line-segment as side to construct a regular hexagon.

Let AB be the given segment.

Constr.—On AB construct the equilateral triangle AOB (124°, Cor. 1), and with O as centre describe a circle through A, cutting AO and BO produced in D and E. Draw FC, the internal bisector of ∠AOE. Then ABCDEF is the hexagon.

Proof.—   ∠AOB=∠EOD=$\frac{2}{3}$⌐,
∴   ∠AOE=$\frac{4}{3}$⌐, and AOF=$\frac{2}{3}$⌐,
∴   ∠AOB=∠BOC=∠COD=etc.=$\frac{2}{3}$⌐.

And the chords AB, BC, CD, etc., being sides of congruent equilateral triangles are all equal.

Therefore ABCDEF is a regular hexagon.

Cor. Since AOB is an equilateral triangle, AB=AO;

∴ the side of a regular hexagon is equal to the radius of its circumcircle.

135°. *Problem.*—To determine which species of regular polygons, each taken alone, can fill the plane.

That a regular polygon of any species may be capable of filling the plane, the number of right angles in its internal angle must be a divisor of 4. But as no internal angle can be so great as two right angles, the only divisors, in 133°, Cor., are $\frac{2}{3}$, 1, and $\frac{4}{3}$, which give the quotients 6, 4, and 3.

Therefore the plane can be filled by 6 equilateral triangles, or 4 squares, or 3 hexagons.

It is worthy of note that, of the three regular polygons which can fill the plane, the hexagon includes the greatest area for a given perimeter. As a consequence, the hexagon is frequently found in Nature, as in the cells of bees, in certain tissues of plants, etc.

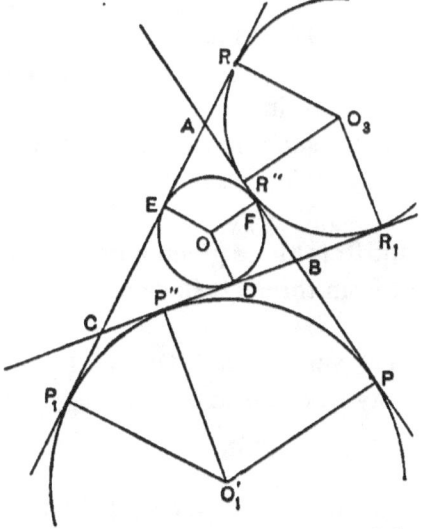

Ex. 1. Let D, E, F be points of contact of the incircle, and P, P', P'', R, R', R'', etc., of the ex-circles.

(131°)

Then AP=AP', CP'=CP'', and BP=BP'', (114°, Cor. 1)

∴ AP'+AP
= AB+BC+AC
= $a+b+c$,

and, denoting the perimeter of the triangle by $2s$, we have

AP=AP'=s,

CONSTRUCTIVE GEOMETRY. 89

$\therefore$ $\quad CP' = s-b = CP''$, $\quad BP = s-c = BP''$.
Similarly, $\quad AR = s-b = AR''$, $\quad BR' = s-a = BR''$, etc.
Again, $\quad CD = CE = b - AE = b - AF = b - (c - BF)$
$\qquad = b - c + BD = b - c + a - CD$,
$\therefore \quad 2CD = b + a - c = 2(s-c)$,
$\therefore \quad CE = CD = s - c = BP''$.
Similarly, $\quad AE = AF = s - a = BR''$, etc.

These relations are frequently useful.

If we put Ai to denote the distance of the vertex A from the adjacent points of contact of the in-circle, and Ab, Ac to denote its distances from the points of contact of the ex-circles upon the sides b and c respectively, we have

$$Ai = Bc = Cb = s - a,$$
$$Bi = Ca = Ac = s - b,$$
$$Ci = Ab = Ba = s - c.$$

Exercises.

1. In testing the straightness of a "rule" three rules are virtually tested. How?
2. To construct a rectangle, and also a square.
3. To place a given line-segment between two given lines so as to be parallel to a given line.
4. On a given line to find a point such that the lines joining it to two given points may make equal angles with the given line.
5. To find a point equidistant from three given points.
6. To find a line equidistant from three given points. How many lines?
7. A is a point on line L and B is not on L. To find a point P such that $PA \pm PB$ may be equal to a given segment.
8. On a given line to find a point equidistant from two given points.
9. Through a given point to draw a line which shall form an isosceles triangle with two given lines. How many solutions?

10. Through two given points on two parallel lines to draw two lines so as to form a rhombus.
11. To construct a square having one of its vertices at a given point, and two other vertices lying on two given parallel lines.
12. Through a given point to draw a line so that the intercept between two given parallels may be of a given length.
13. To construct a triangle when the basal angles and the altitude are given.
14. To construct a right-angled triangle when the hypothenuse and the sum of the sides are given.
15. To divide a line-segment into any number of equal parts.
16. To construct a triangle when the middle points of its sides are given.
17. To construct a parallelogram when the diagonals and one side are given.
18. Through a given point to draw a secant so that the chord intercepted by a given circle may have a given length.
19. Draw a line to touch a given circle and be parallel to a given line. To be perpendicular to a given line.
20. Describe a circle of given radius to touch two given lines.
21. Describe a circle of given radius to touch a given circle and a given line.
22. Describe a circle of given radius to pass through a given point and touch a given circle.
23. Describe a circle of given radius to touch two given circles.
24. To inscribe a regular octagon in a circle.
25. To inscribe a regular dodecagon in a circle.
26. A, B, C, D, ..., are consecutive vertices of a regular octagon, and A, B', C', D', ..., of a regular dodecagon in the same circle. Find the angles between AC and B'C'; between BE' and B'E. (Use 108°.)
27. Show that the plane can be filled by
    (*a*) Equilateral triangles and regular dodecagons.
    (*b*) Equilateral triangles and squares.
    (*c*) Squares and regular octagons.

PART II.

PRELIMINARY.

136°. *Def.* 1.—The *area* of a plane closed figure is the portion of the plane contained within the figure, this portion being considered with respect to its extent only, and without respect to form.

A closed figure of any form may contain an area of any given extent, and closed figures of different forms may contain areas of the same extent, or equal areas.

Def. 2.—Closed figures are *equal* to one another when they include equal areas. This is the definition of the term "equal" when comparing closed figures.

Congruent figures are necessarily equal, but equal figures are not necessarily congruent. Thus, a △ and a ▭ may have equal areas and therefore be *equal*, although necessarily having different forms.

137°. Areas are compared by superposition. If one area can be superimposed upon another so as exactly to cover it, the areas are equal and the figures containing the areas are equal. If such superposition can be shown to be impossible the figures are not equal.

In comparing areas we may suppose one of them to be divided into any requisite number of parts, and these parts to be afterwards disposed in any convenient order, since the whole area is equal to the sum of all its parts.

Illustration.—ABCD is a square.

Then the △ABC ≡ △ADC, and they are therefore equal.

Now, if AD and DE be equal and in line, the △s ADC and EDC are congruent and equal.

Therefore the △ABC may be taken from its present position and be put into the position of CDE. And the square ABCD is thus transformed into the △ACE without any change of area;

∴ □ABCD = △ACE.

It is evident that a plane closed figure may be considered from two points of view.

1. With respect to the character and disposition of the lines which form it. When thus considered, figures group themselves into triangles, squares, circles, etc., where the members of each group, if not of the same form, have at least some community of form and character.

2. With respect to the areas enclosed.

When compared from the first point of view, the capability of superposition is expressed by saying that the figures are congruent. When compared from the second point of view, it is expressed by saying that the figures are equal.

Therefore congruence is a kind of higher or double equality, that is, an equality in both form and extent of area. This is properly indicated by the triple lines (≡) for congruence, and the double lines (=) for equality.

138°. *Def.*—The altitude of a figure is the line-segment which measures the distance of the farthest point of a figure from a side taken as base.

The terms *base* and *altitude* are thus correlative. A triangle may have three different bases and as many corresponding altitudes. (87°)

In the *rectangle* (82°, Def. 2) two adjacent sides being perpendicular to one another, either one may be taken as the base and the adjacent one as the altitude. The rectangle having two given segments as its base and altitude is called the *rectangle on these segments*.

Notation.—The symbol ▭ stands for the word *rectangle* and ▱ for *parallelogram*.

Rectangles and parallelograms are commonly indicated by naming a pair of their opposite vertices.

SECTION I.

COMPARISON OF AREAS—RECTANGLES, PARALLELOGRAMS, TRIANGLES.

139°. *Theorem.*—1. Rectangles with equal bases and equal altitudes are equal.

2. Equal rectangles with equal bases have equal altitudes.
3. Equal rectangles with equal altitudes have equal bases.

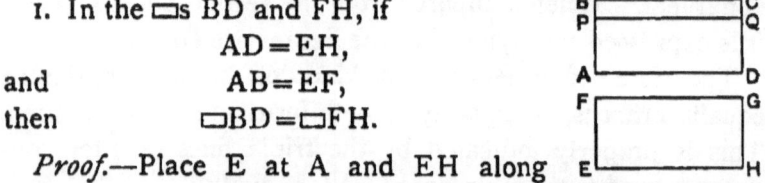

1. In the ▭s BD and FH, if
$$AD = EH,$$
and $AB = EF,$
then $▭BD = ▭FH.$

Proof.—Place E at A and EH along AD. Then, as ∠FEH = ∠BAD = ⌐, EF will lie along AB.

And because EH = AD and EF = AB, therefore H falls at D and F at B, and the two ▭s are congruent and therefore equal. *q.e.d.*

2. If ▭BD = ▭FH and AD = EH, then AB = EF.

Proof.—If EF is not equal to AB, let AB be > EF. Make AP = EF and complete the ▭PD.
Then $▭PD = ▭FH,$ by the first part,
but $▭BD = ▭FH,$ (hyp.)

∴ ▭PD = ▭BD, which is not true,
∴ AB and EF cannot be unequal, or
AB = EF. *q.e.d.*

3. If ▭BD = ▭FH and AB = EF, then AD = EH.

Proof.—Let AB and EF be taken as bases and AD and EH as altitudes (138°), and the theorem follows from the second part. *q.e.d.*

Cor. In any rectangle we have the three parts, base, altitude, and area. If any two of these are given the third is given also.

140°. *Theorem.*—A parallelogram is equal to the rectangle on its base and altitude.

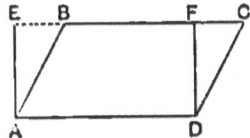

AC is a ▱ whereof AD is the base and DF is the altitude.
Then ▱AC = ▭ on AD and DF.

Proof.—Complete the ▭ADFE by drawing AE ⊥ to CB produced.
Then △AEB ≡ △DFC, ∵ AE = DF, AB = DC,
and ∠EAB = ∠FDC;
∴ △DFC may be transferred to the position AEB, and ▱ABCD becomes the ▭AEFD,
∴ ▱AC = ▭ on AD and DF. *q.e.d.*

Cor. 1. Parallelograms with equal bases and equal altitudes are equal. For they are equal to the same rectangle.

Cor. 2. Equal parallelograms with equal bases have equal altitudes, and equal parallelograms with equal altitudes have equal bases.

Cor. 3. If equal parallelograms be upon the same side of the same base, their sides opposite the common base are in line.

141°. *Theorem.*—A triangle is equal to one-half the rectangle on its base and altitude.

COMPARISON OF AREAS.

ABC is a triangle of which AC is the base and BE the altitude.

Then △ABC = ½▱ on AC and BE.

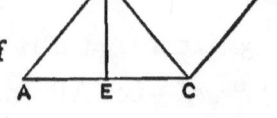

Proof.—Complete the ▱ABDC, of which AB and AC are adjacent sides.
Then △ABC ≡ △DCB,
∴ △ABC = ½▱AD = ½▱ on AC and BE. (140°) *q.e.d.*

Cor. 1. A triangle is equal to one-half the parallelogram having the same base and altitude.

Cor. 2. Triangles with equal bases and equal altitudes are equal. For they are equal to one-half of the same rectangle.

Cor. 3. A median of a triangle bisects the area. For the median bisects the base.

Cor. 4. Equal triangles with equal bases have equal altitudes, and equal triangles with equal altitudes have equal bases.

Cor. 5. If equal triangles be upon the same side of the same base, the line through their vertices is parallel to their common base.

142°. *Theorem.*— If two triangles are upon opposite sides of the same base—

1. When the triangles are equal, the base bisects the segment joining their vertices;

2. When the base bisects the segment joining their vertices, the triangles are equal. (Converse of 1.)

ABC and ADC are two triangles upon opposite sides of the common base AC.

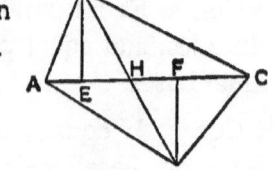

1. If △ABC = △ADC,
then BH = HD.

Proof.—Let BE and DF be altitudes,
Then ∵ △ABC = △ADC, ∴ BE = DF,
∴ △EBH ≡ △FDH, and BH = HD. *q.e.d.*

2. If BH = HD, then △ABC = △ADC.

Proof.—Since BH = HD, ∴ △ABH = △ADH,
and △CBH = △CDH. (141°, Cor. 3)
∴ adding, △ABC = △ADC. *q.e.d.*

143. *Def.*—By the *sum* or *difference* of two closed figures is meant the sum or difference of the areas of the figures.

If a rectangle be equal to the sum of two other rectangles its area may be so superimposed upon the others as to cover both.

144°. *Theorem.*—If two rectangles have equal altitudes, their sum is equal to the rectangle on their common altitude and the sum of their bases.

Proof.—Let the ▭s X and Y, having equal altitudes, be so placed as to have their altitudes in common at CD, and so that one ▭ may not overlap the other.

Then ∠BDC = ∠CDF = ⌐,
∴ BDF is a line. (38°, Cor. 2)
Similarly ACE is a line.

But BD is ∥ to AC, and BA is ∥ to DC ∥ to FE; therefore AF is the ▭ on the altitude AB and the sum of the bases AC and CE; and the ▭AF = ▭AD + ▭CF. *q.e.d.*

Cor. 1. If two triangles have equal altitudes, their sum is equal to the triangle having the same altitude and having a base equal to the sum of the bases of the two triangles.

Cor. 2. If two triangles have equal altitudes, their sum is equal to one-half the rectangle on their common altitude and the sum of their bases.

Cor. 3. If any number of triangles have equal altitudes, their sum is equal to one-half the rectangle on their common altitude and the sum of their bases.

In any of the above, "base" and "altitude" are interchangeable.

COMPARISON OF AREAS. 97

145°. *Theorem.*—Two lines parallel to the sides of a parallelogram and intersecting upon a diagonal divide the parallelogram into four parallelograms such that the two through which the diagonal does not pass are equal to one another.

In the ▱ABCD, EF is ∥ to AD and GH is ∥ to BA, and these intersect at O on the diagonal AC.

Then ▱BO = ▱OD.

Proof.—△ABC = △ADC, and △AEO = △AHO,
and △OGC = △OFC ; (141°, Cor. 1)
but ▱BO = △ABC − △AEO − △OGC,
and ▱OD = △ADC − △AHO − △OFC.
∴ ▱BO = ▱OD. *q.e.d.*

Cor. 1. ▱BF = ▱GD.

Cor. 2. If ▱BO = ▱OD, O is on the diagonal AC. (Converse of the theorem.)

For if O is not on the diagonal, let the diagonal cut EF in O'. Then ▱BO' = ▱O'D. (145°)

But ▱BO' is < ▱BO, and ▱O'D is > ▱OD ;
∴ ▱BO is > ▱OD, which is contrary to the hypothesis;
∴ the diagonal cuts EF in O.

Ex. Let ABCD be a trapezoid. (84°, Def.) In line with AD make DE = BC, and in line with BC make CF = AD.

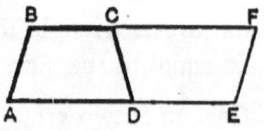

Then BF = AE and BFEA is a ▱.

But the trapezoid CE can be superimposed on the trapezoid DB, since the sides are respectively equal, and

∠F = A, and ∠E = B, etc.

∴ trapezoid BD = ½ ▱BE,

or, a trapezoid is equal to one-half the rectangle on its altitude and the sum of its bases.

Exercises.

1. To construct a triangle equal to a given quadrangle.
2. To construct a triangle equal to a given polygon.
3. To bisect a triangle by a line drawn through a given point in one of the sides.
4. To construct a rhombus equal to a given parallelogram, and with one of the sides of the parallelogram as its side.
5. The three connectors of the middle points of the sides of a triangle divide the triangle into four equal triangles.
6. Any line concurrent with the diagonals of a parallelogram bisects the parallelogram.
7. The triangle having one of the non-parallel sides of a trapezoid as base and the middle point of the opposite side as vertex is one-half the trapezoid.
8. The connector of the middle points of the diagonals of a quadrangle is concurrent with the connectors of the middle points of opposite sides.
9. ABCD is a parallelogram and O is a point within. Then
$$\triangle AOB + \triangle COD = \tfrac{1}{2}\square.$$
What does this become when O is without?
10. ABCD is a parallelogram and O is a point within. Then
$$\triangle AOC = \triangle AOD - \triangle AOB.$$
What does this become when O is without? (This theorem is important in the theory of Statics.)
11. Bisect a trapezoid by a line through the middle point of one of the parallel sides. By a line through the middle point of one of the non-parallel sides.
12. The triangle having the three medians of another triangle as its sides has three-fourths the area of the other.

POLYGON AND CIRCLE.

146°. *Def.*—The sum of all the sides of a polygon is called its *perimeter*, and when the polygon is regular every side is at the same distance from the centre. This distance is the *apothem* of the polygon.

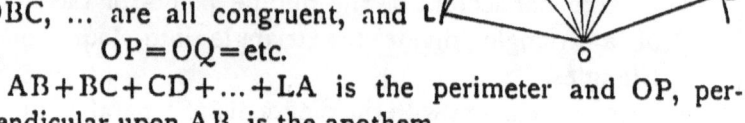

Thus if ABCD...LA be a regular polygon and O the centre (132°, Def. 2), the triangles OAB, OBC, ... are all congruent, and
∴ OP = OQ = etc.

AB + BC + CD + ... + LA is the perimeter and OP, perpendicular upon AB, is the apothem.

147°. *Theorem.*—A regular polygon is equal to one-half the rectangle on its apothem and perimeter.

Proof.—The triangles AOB, BOC, ... LOA have equal altitudes, the apothem OP, ∴ their sum is one-half the ▭ on OP and the sum of their bases AB + BC + ...LA. (144°, Cor. 3)

But the sum of the triangles is the polygon, and the sum of their bases is the perimeter.

∴ a regular polygon = ½▭ on its apothem and perimeter.

148°. *Of a limit.*—A *limit* or *limiting value* of a variable is the value to which the variable by its variation can be made to approach indefinitely near, but which it can never be made to pass.

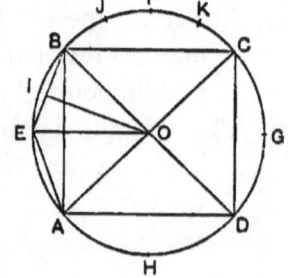

Let ABCD be a square in its circumcircle. If we bisect the arcs AB, BC, CD, and DA in E, F, G, and H, we have the vertices of a regular octagon AEBFCGDHA. Now, the area of the octagon approaches nearer to that of the circle than the area of the square does; and the perimeter of the octagon approaches nearer to the length of the circle than the perimeter of the square does; and the apothem of the octagon approaches nearer to the radius of the circle than the apothem of the square does.

Again, bisecting the arcs AE, EB, BF, etc., in I, J, K, etc., we obtain the regular polygon of 16 sides. And all the foregoing parts of the polygon of 16 sides approach nearer to the corresponding parts of the circle than those of the octagon do.

It is evident that by continually bisecting the arcs, we may obtain a series of regular polygons, of which the last one may be made to approach the circle as near as we please, but that however far this process is carried the final polygon can never become greater than the circle, nor can the final apothem become greater than the radius.

Hence the circle is the limit of the perimeter of the regular polygon when the number of its sides is endlessly increased, and the area of the circle is the limit of the area of the polygon, and the radius of the circle is the limit of the apothem of the polygon under the same circumstances.

149°. *Theorem.*—A circle is equal to one-half the rectangle on its radius and a line-segment equal in length to the circle.

Proof.—The ⊙ is the limit of a regular polygon when the number of its sides is endlessly increased, and the radius of the ⊙ is the limit of the apothem of the polygon.

But, whatever be the number of its sides, a regular polygon is equal to one-half the ▭ on its apothem and perimeter. (147°)

∴ a ⊙ is equal to one-half the ▭ on its radius and a line-segment equal to its circumference.

Exercises.

1. Show that a regular polygon may be described about a circle, and that the limit of its perimeter when the number of its sides is increased indefinitely is the circumference of the circle.

2. The difference between the areas of two regular polygons, one inscribed in a circle and the other circumscribed about it, vanishes at the limit when the number of sides of the polygons increases indefinitely.

3. What is the limit of the internal angle of a regular polygon as the number of its sides is endlessly increased?

SECTION II.

MEASUREMENT OF LENGTHS AND AREAS.

150°. *Def.*—1. That part of Geometry which deals with the measures and measuring of magnitudes is *Metrical* Geometry.

2. To measure a magnitude is to determine how many unit magnitudes of the same kind must be taken together to form the given magnitude. And the number thus determined is called the *measure* of the given magnitude with reference to the unit employed. This number may be a whole or a fractional number, or a numerical quantity which is not arithmetically expressible. The word "number" will mean any of these.

3. In measuring length, such as that of a line-segment, the unit is a segment of arbitrary length called the *unit-length*. In practical work we have several such units as an inch, a foot, a mile, a metre, etc., but in the Science of Geometry the unit-length is quite arbitrary, and results obtained through it are so expressed as to be independent of the length of the particular unit employed.

4. In measuring areas the unit magnitude is the area of the square having the unit-length as its side. This area is the *unit-area*. Hence the unit-length and unit-area are not *both*

arbitrary, for if either is fixed the other is fixed also, and determinable.

This relation between the unit-length and the unit-area is conventional, for we might assume the unit-area to be the area of any figure which is wholly determined by a single segment taken as the unit-length: as, for example, an equilateral triangle with the unit-length as side, a circle with the unit-length as diameter, etc. The square is chosen because it offers decided advantages over every other figure.

For the sake of conciseness we shall symbolize the term unit-length by *u.l.* and unit-area by *u.a.*

5. When two magnitudes are such that they are both capable of being expressed arithmetically in terms of some common unit they are *commensurable*, and when this is not the case they are *incommensurable*.

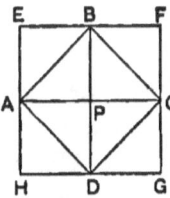

Illus.—Let ABCD be a square, and let EF and HG be drawn ⊥ to BD, and EH and FG ⊥ to AC. Then EFGH is a square (82°, Cor. 5), and the triangles AEB, APB, BFC, BPC, etc., are all equal to one another.

If AB be taken as *u.l.*, the area of the square AC is the *u.a.*; and if EF be taken as *u.l.*, the area of the square EG is the *u.a.*

In the first case the measure of the square AC is 1, and that of EG is 2; and in the latter case the measure of the square EG is 1, and that of AC is $\frac{1}{2}$. So that in both cases the measure of the square EG is double that of the square AC.

∴ the squares EG and AC are commensurable.

Now, if AB be taken as *u.l.*, EF is not expressible arithmetically, as will be shown hereafter.

∴ AB and EF are incommensurable.

151°. Let AB be a segment trisected at E and F (127°), and let AC be the square on AB. Then AD=AB. And if AD be trisected in the points K and M, and through E and F ‖s be drawn to AD, and through K and M ‖s be drawn to AB, the figures 1, 2, 3, 4, 5, 6, 7, 8, 9 are all squares equal to one another.

Now, if AB be taken as *u.l.*, AC is the *u.a.*; and if AE be taken as *u.l.*, any one of the small squares, as AP, is the *u.a.* And the segment AB contains AE 3 times, while the square AC contains the square AP in three rows with three in each row, or 3^2 times.

∴ if any assumed *u.l.* be divided into 3 equal parts for a new *u.l.*, the corresponding *u.a.* is divided into 3^2 equal parts for a new *u.a.* And the least consideration will show that this is true for any whole number as well as 3.

∴ 1. If an assumed *u.l.* be divided into n equal parts for a new *u.l.*, the corresponding *u.a.* is divided into n^2 equal parts for a new *u.a.*; n denoting any whole number.

Again, if any segment be measured by the *u.l.* AB, and also by the *u.l.* AE, the measure of the segment in the latter case is three times that in the former case. And if any area be measured by the *u.a.* AC, and also by the *u.a.* AP, the measure of the area in the latter case is 3^2 times its measure in the former case. And as the same relations are evidently true for any whole number as well as 3,

∴ 2. If any segment be measured by an assumed *u.l.* and also by $\frac{1}{n}$th of the assumed *u.l.* as a new *u.l.*, the measure of the segment in the latter case is n times its measure in the former. And if any area be measured by the corresponding *u.a.*s the measure of the area in the latter case is n^2 times its measure in the former case; n being any whole number.

This may be stated otherwise as follows:—

By reducing an assumed *u.l.* to $\frac{1}{n}$th of its original length, we increase the measure of any given segment n times, and we increase the measure of any given area n^2 times; n being a whole number.

In all cases where a *u.l.* and a *u.a.* are considered together, they are supposed to be connected by the relation of 150°, 3 and 4.

152°. *Theorem.*—The number of unit-areas in a rectangle is the product of the numbers of unit-lengths in two adjacent sides.

The proof is divided into three cases.

1. Let the measures of the adjacent sides with respect to the unit adopted be whole numbers.

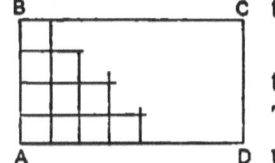

Let AB contain the assumed *u.l.* a times, and let AD contain it b times. Then, by dividing AB into a equal parts and drawing, through each point of division, lines ∥ to AD, and by dividing AD into b equal parts and drawing, through each point of division, lines ∥ to AB, we divide the whole rectangle into equal squares, of which there are a rows with b squares in each row.

∴ the whole number of squares is ab.

But each square has the *u.l.* as its side and is therefore the *u.a.*

∴ *u.a.*s in AC = *u.l.*s in AB × *u.l.*s in AD.

We express this relation more concisely by writing symbolically ▭AC = AB . AD,

where ▭AC means "the number of *u.a.*s in ▭AC," and AB and AD mean respectively "the numbers of *u.l.*s in these sides."

And in language we say, the area of a rectangle is the product of its adjacent sides; the proper interpretation of which is easily given.

2. Let the measures of the adjacent sides with respect to the unit adopted be fractional.

Then, ∵ AB and AD are commensurable, some unit will be an aliquot part of each (150°, 5). Let the new unit be $\frac{1}{n}$th of the adopted unit, and let AB contain p of the new units, and AD contain q of them.

The measure of ▭AC in terms of the new *u.a.* is pq

(152°, 1), and the measure of the ☐AC in terms of the adopted unit is $\frac{pq}{n^2}$ (151°, 2)

But the measure of AB in terms of the adopted *u.l.* is $\frac{p}{n}$, and of AD it is $\frac{q}{n}$. (151°, 2)

and $$\frac{pq}{n^2} = \frac{p}{n} \cdot \frac{q}{n},$$

or ☐AC = AB . AD.

Illus.—Suppose the measures of AB and AD to some unit-length to be 3.472 and 4.631. By taking a *u.l.* 1000 times smaller these measures become the whole numbers 3472 and 4631, and the number of corresponding *u.a.*s in the rectangle is 3472 × 4631 or 16078832;
and dividing by 1000², the measure of the area with respect to the original *u.l.* is 16.078832 = 3.472 × 4.631.

3. Let the adjacent sides be incommensurable. There is now no *u.l.* that will measure both AB and AD.

If ☐AC is not equal to AB . AD, let it be equal to AB . AE, where AE has a measure different from AD; and suppose, first, that AE is < AD, so that E lies between A and D.

With any *u.l.* which will measure AB, and which is less than ED, divide AD into parts. One point of division at least must fall between E and D; let it fall at H. Complete the rectangle BH.

Then AB and AH are commensurable, and
☐BH = AB . AH,
but ☐BD = AB . AE; (hyp.)
and ☐BH is < ☐BD;
∴ AB . AH is < AB . AE,
and AB being a common factor
AH is < AE; which is not true.

∴ If ☐AC = AB . AE, AE cannot be < AD, and similarly it may be shown that AE cannot be > AD; ∴ AE = AD, or
☐AC = AB . AD. *q.e.d.*

153°. The results of the last article in conjunction with Section I. of this Part give us the following theorems.

1. The area of a parallelogram is the product of its base and altitude. (140°)

2. The area of a triangle is one-half the product of its base and altitude. (141°)

3. The area of a trapezoid is one-half the product of its altitude and the sum of its parallel sides. (145°, Ex.)

4. The area of any regular polygon is one-half the product of its apothem and perimeter. (147°)

5. The area of a circle is one-half the product of its radius and a line-segment equal to its circumference. (149°)

Ex. 1. Let O, O' be the centres of the in-circle and of the ex-circle to the side BC (131°); and let OD, O'P'' be perpendiculars on BC, OE, O'P' perpendiculars on AC, and OF, O'P on AB. Then
$$OD = OE = OF = r$$
and $O'P = O'P' = O'P'' = r'$;

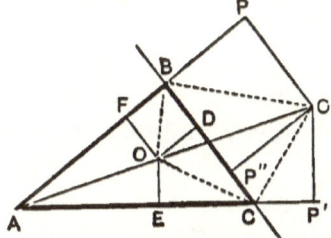

$$\therefore \triangle ABC = \triangle AOB + \triangle BOC + \triangle COA$$
$$= \tfrac{1}{2}AB \cdot OF + \tfrac{1}{2}BC \cdot OD + \tfrac{1}{2}CA \cdot OE \qquad (153°, 2)$$
$$= \tfrac{1}{2}r \times \text{perimeter} = rs,$$

where s is the half perimeter;

$$\therefore \triangle = rs.$$

Ex. 2. $\triangle ABC = \triangle AO'B + \triangle AO'C - \triangle BO'C$
$$= \tfrac{1}{2}O'P \cdot AB + \tfrac{1}{2}O'P' \cdot AC - \tfrac{1}{2}O'P'' \cdot BC$$
$$= \tfrac{1}{2}r'(b + c - a) = r'(s - a),$$

where r' is the radius of the ex-circle to side a;

$$\therefore \triangle = r'(s - a).$$

Similarly, $\triangle = r''(s - b)$
$$= r'''(s - c).$$

Exercises.

1. $\dfrac{1}{r} = \dfrac{1}{r'} + \dfrac{1}{r''} + \dfrac{1}{r'''}.$

2. $\Delta^2 = rr'r''r'''.$

3. What relation holds between the radius of the in-circle and that of an ex-circle when the triangle is equiangular?

Note.—When the diameter of a circle is taken as the *u.l.* the measure of the circumference is the inexpressible numerical quantity symbolized by the letter π, and which, expressed approximately, is 3.1415926....

4. What is the area of a square when its diagonal is taken as the *u.l.*?

5. What is the measure of the diagonal of a square when the side is taken as the *u.l.*? (150°, 5)

6. Find the measure of the area of a circle when the diameter is the *u.l.* When the circumference is the *u.l.*

7. If one line-segment be twice as long as another, the square on the first has four times the area of the square on the second. (151°, 2)

8. If one line-segment be twice as long as another, the equilateral triangle on the first is four times that on the second. (141°)

9. The equilateral triangle on the altitude of another equilateral triangle has an area three-fourths that of the other.

10. The three medians of any triangle divide its area into six equal triangles.

11. From the centroid of a triangle draw three lines to the sides so as to divide the triangle into three equal quadrangles.

12. In the triangle ABC X is taken in BC, Y in CA, and Z in AB, so that BX=⅓BC, CY=⅓CA, and AZ=⅓AB. Express the area of the triangle XYZ in terms of that of ABC.

13. Generalize 12 by making $BX = \dfrac{1}{n} BC$, etc.

14. Show that $a = s\left(1 - \dfrac{r}{r'}\right) = \dfrac{r'}{s}(r'' + r''').$

SECTION III.

GEOMETRIC INTERPRETATION OF ALGEBRAIC FORMS.

154°. We have a language of symbols by which to express and develop mathematical relations, namely, Algebra. The symbols of Algebra are quantitative and operative, and it is very desirable, while giving a geometric meaning to the symbol of quantity, to so modify the meanings of the symbols of operation as to apply algebraic forms in Geometry. This application shortens and generalizes the statements of geometric relations without interfering with their accuracy.

Elementary Algebra being generalized Arithmetic, its quantitative symbols denote numbers and its operative symbols are so defined as to be consistent with the common properties of numbers.

Thus, because $2+3=3+2$ and $2.3=3.2$, we say that $a+b=b+a$ and $ab=ba$.

This is called the *commutative* law. The first example is of the existence of the law in addition, and the second of its existence in multiplication.

The commutative law in addition may be thus expressed:— A sum is independent of the order of its addends; and in multiplication—A product is independent of the order of its factors.

Again, because $2(3+4)=2.3+2.4$, we say that
$$a(b+c)=ab+ac.$$

This is called the *distributive* law and may be stated thus:—The product of multiplying a factor by the sum of several terms is equal to the sum of the products arising from multiplying the factor by each of the terms.

These two are the only laws which need be here mentioned. And any science which is to employ the *forms* of Algebra

INTERPRETATION OF ALGEBRAIC FORMS. 109

must have that, whatever it may be, which is denoted by the algebraic symbol of quantity, subject to these laws.

155°. As already explained in 22° we denote a single line-segment, in the one-letter notation, by a single letter, as a, which is equivalent to the algebraic symbol of quantity; and hence,

A single algebraic symbol of quantity is to be interpreted geometrically as a line-segment.

It must of course be understood, in all cases, that in employing the two-letter notation for a segment (22°), as "AB," the two letters standing for a single line-segment are equivalent to but a single algebraic symbol of quantity.

The expression $a+b$ denotes a segment equal in length to those denoted by a and b together.

Similarly $2a=a+a$, and na means a segment as long as n of the segments a placed together in line, n being any numerical quantity whatever. (28°)

$a-b$, when a is longer than b, is the segment which is left when a segment equal to b is taken from a.

Now it is manifest that, if a and b denote two segments, $a+b=b+a$, and hence that the commutative law for addition applies to these symbols when they denote magnitudes having length only, as well as when they denote numbers.

156°. *Line in Opposite Senses.*—A quantitative symbol, a, is in Algebra always affected with one of two signs, $+$ or $-$, which, while leaving the absolute value of the symbol unchanged, impart to it certain properties exactly opposite in character.

This oppositeness of character finds its complete interpretation in Geometry in the opposite directions of every segment. Thus the segment in the margin may be considered as extending *from* A to B or *from* B to A.

With the two-letter notation the direction can be denoted by the order of the letters, and this is one of the advantages of this notation; but with the one-letter notation, if we denote the segment AB by $+a$, we *must* denote the segment BA by $-a$.

But as there is no absolute reason why one direction rather than the other should be considered positive, we express the matter by saying that AB and BA, or $+a$ and $-a$, denote the same segment taken in *opposite senses*.

Hence the algebraic distinction of positive and negative as applied to a single symbol of quantity is to be interpreted geometrically by the oppositeness of direction of the segment denoted by the symbol.

Usually the applications of this principle in Geometry are confined to those cases in which the segments compared as to sign are parts of one and the same line or are parallel.

Ex. 1. Let ABC be any $\triangle$ and let BD be the altitude from the vertex B.

Now, suppose that the sides AB and BC undergo a gradual change, so that B may move along the line BB' until it comes into the position denoted by B'.

Then the segment AD gradually diminishes as D approaches A; disappears when D coincides with A, in which case B comes to be vertically over A and the $\triangle$ becomes right-angled at A; reappears as D passes to the left of A, until finally we may suppose that one stage of the change is represented by the $\triangle$AB'C with its altitude B'D'.

Then, if we call AD positive, we *must* call AD' negative, or we must consider AD and AD' as having opposite senses.

Again, from the principle of continuity (104°) the foot of the altitude cannot pass from D on the right of A to D' on the left of A without passing through every intermediate point, and therefore passing *through* A. And thus the segment AD must vanish before it changes sign.

This is conveniently expressed by saying that *a line-*

segment changes sign when it passes through zero; passing through zero being interpreted as vanishing and reappearing on the other side of the zero-point.

Ex. 2. ABCD is a normal quadrangle. Consider the side AD and suppose D to move along the line DA until it comes into the position D'.

The segments AD and AD' are opposite in sense, and ABCD' is a crossed quadrangle.

∴ the crossed quadrangle is derived from the normal one by changing the sense of one of the sides.

Similarly, if one of the sides of a crossed quadrangle be changed in sense the figure ceases to be a crossed quadrangle.

Ex. 3. This is an example where segments which are parallel but which are not in line have opposite senses.

ABC is a △ and P is any point within from which perpendiculars PD, PE, PF are drawn to the sides.

Suppose that P moves to P'. Then PF becomes P'F', and PF and P'F' being in the same direction have the same sense. Similarly PE becomes P'E', and these segments have the same sense. But PD becomes P'D' which is read in a direction opposite to that of PD. Hence PD and P'D' are opposite in sense.

But PD and P'D' are perpendiculars to the same line from points upon opposite sides of it, and it is readily seen that in passing from P to P' the ⊥PD becomes zero and then changes sense as P crosses the side BC.

Hence if by any continuous change in a figure a point passes from one side of a line to the other side, the perpendicular from that point to the line changes sense.

Cor. If ABC be equilateral it is easily shown that
$$PD + PE + PF = \text{a constant}.$$

And if we regard the sense of the segments this statement is true for all positions of P in the plane.

157°. *Product*.—The algebraic form of a product of two symbols of quantity is interpreted geometrically by the rectangle having for adjacent sides the segments denoted by the quantitative symbols.

This is manifest from Art. 152°, for in the form ab the single letters may stand for the measures of the sides, and the product ab will then be the measure of the area of the rectangle.

If we consider ab as denoting a $\square$ having a as altitude and b as base, then ba will denote the $\square$ having b as altitude and a as base. But in any $\square$ it is immaterial which side is taken as base (138°); therefore $ab = ba$, and the form satisfies the commutative law for multiplication.

Again, let AC be the segment $b + c$, and AB be the segment

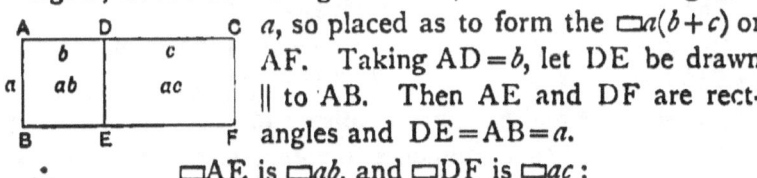

a, so placed as to form the $\square a(b+c)$ or AF. Taking AD $= b$, let DE be drawn $\parallel$ to AB. Then AE and DF are rectangles and DE $=$ AB $= a$.

∴ $\square$AE is $\square ab$, and $\square$DF is $\square ac$;

∴ $\square a(b+c) = \square ab + \square ac$,

and the distributive law is satisfied.

158°. We have then the two following interpretations to which the laws of operation of numbers apply whenever such operations are interpretable.

1. *A single symbol of quantity denotes a line-segment.*

As the sum or difference of two line-segments is a segment, the sum of any number of segments taken in either sense is a segment.

Therefore any number of single symbols of quantity connected by $+$ and $-$ signs denotes a segment, as $a + b$, $a - b + c$, $a - b + (-c)$, etc.

For this reason such expressions or forms are often called *linear*, even in Algebra.

Other forms of linear expressions will appear hereafter.

2. *The product form of two symbols of quantity denotes the rectangle whose adjacent sides are the segments denoted by the single symbols.*

A rectangle encloses a portion of the plane and admits of measures in two directions perpendicular to one another, hence the area of a rectangle is said to be of two dimensions. And as all areas can be expressed as rectangles, areas in general are of two dimensions.

Hence algebraic terms which denote rectangles, such as ab, $(a+b)c$, $(a+b)(c+d)$, etc., are often called rectangular terms, and are said to be of two dimensions.

Ex. Take the algebraic identity
$$a(b+c) = ab + ac.$$
The geometric interpretation gives—

If there be any three segments (a, b, c) the ▭ on the first and the sum of the other two (b, c) is equal to the sum of the ▭s on the first and each of the other two.

The truth of this geometric theorem is evident from an inspection of a proper figure.

This is substantially *Euclid*, Book II., Prop. 1.

159°. *Square.*—When the segment b is equal to the segment a the rectangle becomes the square on a. When this equality of symbols takes place in Algebra we write a^2 for aa, and we call the result the "square" of a, the term "square" being derived from Geometry.

Hence the algebraic form of a square is interpreted geometrically by the square which has for its side the segment denoted by the root symbol.

Ex. In the preceding example let b become equal to a, and
$$a(a+c) = a^2 + ac,$$
H

which interpreted geometrically gives—

If a segment $(a+c)$ be divided into two parts (a, c), the rectangle on the segment and one of its parts (a) is equal to the sum of the square on that part (a^2) and the rectangle on the two parts (ac).

This is *Euclid*, Book II., Prop. 3. The truth of the geometric theorem is manifest from a proper figure.

160°. *Homogeneity.*—Let a, b, c, d denote segments. In the linear expressions $a+b$, $a-b$, etc., and in the rectangular expressions $ab+cd$, etc., the interpretations of the symbols $+$ and $-$ are given in 28°, 29°, and 143°, and are readily intelligible.

But in an expression such as $ab+c$ we have no interpretation for the symbol $+$ if the quantitative symbols denote line-segments. For ab denotes the area of a rectangle and c denotes a segment, and the adding of these is not intelligible in any sense in which we use the word "add."

Hence an expression such as $ab+c$ is not interpretable geometrically This is expressed by saying that—An algebraic form has no geometric interpretation unless the form is *homogeneous*, *i.e.*, unless each of its terms denotes a geometric element of the same kind.

It will be observed that the terms "square," "dimensions," "homogeneous," and some others have been introduced into Algebra from Geometry.

161°. *Rectangles in Opposite Senses.*—The algebraic term ab changes sign if one of its factors changes sign. And to be consistent we must hold that a rectangle changes sense whenever one of its adjacent sides changes sense.

Thus the rectangles $AB.CD$ and $AB.DC$ are the same in extent of area, but have opposite senses. And

$$AB.CD + AB.DC = 0,$$

for the sum $= AB(CD + DC),$

and $CD + DC = 0.$ (156°)

INTERPRETATION OF ALGEBRAIC FORMS.   115

As the sense of a rectangle depends upon that of a line-segment there is no difficulty in determining when rectangles are to be taken in different senses.

The following will illustrate this part of the subject :—

Let OA = OA' and OC = OC', and let the figures be rectangles.

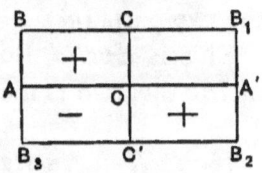

▭s OA.OC and OA'.OC have the common altitude OC and bases equal in length but opposite in sense. Therefore OA.OC and OA'.OC are opposite in sense, and if we call ▭OA.OC positive we must call ▭OA'.OC negative.

Again, ▭s OC.OA' and OC'.OA' have the common base OA' and altitudes equal in length but opposite in sense. Therefore ▭s OC.OA' and OC'.OA' are opposite in sense, and therefore ▭s OA.OC and OA'.OC' are of the same sense. Similarly ▭s OC.OA' and OC'.OA are of the same sense.

These four ▭s are equivalent to the algebraic forms :—

$$+a.+b = +ab, \qquad -a.+b = -ab,$$
$$+a.-b = -ab, \qquad -a.-b = +ab.$$

Ex. 1. ABCD is a normal quadrangle whose opposite sides meet in O, and OE, OF are altitudes of the △s DOC and AOB respectively.

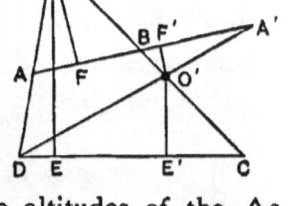

The Qd. ABCD
$= \triangle DOC - \triangle AOB$,
$= \tfrac{1}{2}\square DC.OE - \tfrac{1}{2}\square AB.OF.$ (141°)

Now, let A move along AB to A' (104°). Then O comes to O', F to F', E to E', and O'E', O'F' become the altitudes of the △s DO'C and A'O'B respectively.

But O'E' and OE have the same sense, therefore DC.OE and DC.OE' have the same sense.

Also, A'B is opposite in sense to AB, and O'F' is opposite in sense to OF. (156°, Ex. 3)

∴   AB.OF and A'B.O'F' have the same sense ;
∴   Qd. A'BCD = △DO'C - △A'O'B ;

or, the area of a crossed quadrangle must be taken to be the difference between the two triangles which constitute it.

162°. *Theorem.*—A quadrangle is equal to one-half the parallelogram on its diagonals taken in both magnitude and relative direction.

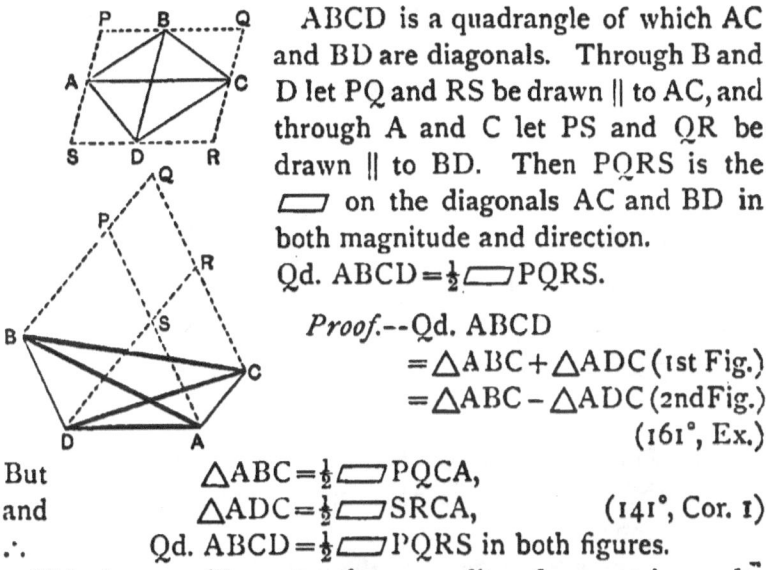

ABCD is a quadrangle of which AC and BD are diagonals. Through B and D let PQ and RS be drawn ∥ to AC, and through A and C let PS and QR be drawn ∥ to BD. Then PQRS is the ▱ on the diagonals AC and BD in both magnitude and direction.
Qd. ABCD = ½ ▱ PQRS.

Proof.—Qd. ABCD
= △ABC + △ADC (1st Fig.)
= △ABC − △ADC (2nd Fig.)
(161°, Ex.)

But △ABC = ½ ▱ PQCA,
and △ADC = ½ ▱ SRCA, (141°, Cor. 1)
∴ Qd. ABCD = ½ ▱ PQRS in both figures.

This theorem illustrates the generality of geometric results when the principle of continuity is observed, and segments and rectangles are considered with regard to sense. Thus the principle of continuity shows that the crossed quadrangle is derived from the normal one (156°, Ex. 2) by changing the sense of one of the sides.

This requires us to give a certain interpretation to the area of a crossed quadrangle (161°, Ex. 1), and thence the present example shows us that all quadrangles admit of a common expression for their areas.

163°. A rectangle is constructed upon two segments which are independent of one another in both length and sense. But a square is constructed upon a single segment, by using

INTERPRETATION OF ALGEBRAIC FORMS.

it for each side. In other words, a rectangle depends upon two segments while a square depends upon only one.

Hence a square can have only one sign, and this is the one which we agree to call positive.

Hence *a square is always positive.*

164°. The algebraic equation $ab = cd$ tells us geometrically that the rectangle on the segments a and b is equal to the rectangle on the segments c and d.

But the same relation is expressed algebraically by the form

$$a = \frac{cd}{b},$$

therefore, since a is a segment, the form $\frac{cd}{b}$ is linear and denotes that segment which with a determines a rectangle equal to cd.

Hence an expression such as $\frac{ab}{c} + \frac{bc}{a} + \frac{ca}{b}$ is linear.

165°. The expression $a^2 = bc$ tells us geometrically that the square whose side is a is equal to the rectangle on the segments b and c.

But this may be changed to the form

$$a = \sqrt{bc}.$$

Therefore since a is a segment, the side of the square, the form $\sqrt{bc}$ is linear.

Hence *the algebraic form of the square root of the product of two symbols of quantity is interpreted geometrically by the side of the square which is equal to the rectangle on the segments denoted by the quantitative symbols.*

166°. The following theorems are but geometric interpretations of well-known algebraic identities. They may, however, be all proved most readily by superposition of areas, and thus the algebraic identity may be derived from the geometric theorem.

1. The square on the sum of two segments is equal to the sum of the squares on the segments and twice the rectangle on the segments.
$$(a+b)^2 = a^2 + b^2 + 2ab.$$

2. The rectangle on the sum and difference of two segments is equal to the difference of the squares on these segments.
$$(a+b)(a-b) = a^2 - b^2.$$

3. The sum of the squares on the sum and on the difference of two segments is equal to twice the sum of the squares on the segments.
$$(a+b)^2 + (a-b)^2 = 2(a^2+b^2), \ a > b.$$

4. The difference of the squares on the sum and on the difference of two segments is equal to four times the rectangle on the segments.
$$(a+b)^2 - (a-b)^2 = 4ab, \ a > b.$$

EXERCISES.

1. To prove 4 of Art. 166°.

Let $AH = a$ and $HB = b$ be the segments, so that AB is their sum. Through H draw HG ∥ to BC, a side of the square on AB. Make $HG = a$, and complete the square FGLE, as in the figure, so that FG is $a - b$.

Then AC is $(a+b)^2$ and EG is $(a-b)^2$; and their difference is the four rectangles AF, HK, CL, and DE; but these each have a and b as adjacent sides.

∴ $(a+b)^2 - (a-b)^2 = 4ab.$

2. State and prove geometrically $(a-b)^2 = a^2 + b^2 - 2ab$.

3. State and prove geometrically
$$(a+b)(a+c) = a^2 + a(b+c) + bc.$$

4. State and prove geometrically by superposition of areas

$(a+b)^2 + (a-b)^2 + 2(a+b)(a-b) = (2a)^2$, where a and b denote segments.

5. If a given segment be divided into any three parts the square on the segment is equal to the sum of the squares on the parts together with twice the sum of the rectangles on the parts taken two and two.

6. Prove, by comparison of areas from the Fig. of Ex. 1, that $(a+b)^2 = 2b(a+b) + 2b(a-b) + (a-b)^2$, and state the theorem in words.

SECTION IV.

AREAL RELATIONS.

167°. *Def.*—1. The segment which joins two given points is called the *join* of the points; and where no reference is made to length the *join* of two points may be taken to mean the line determined by the points.

2. The foot of the perpendicular from a given point to a given line is the *orthogonal projection*, or simply *the projection*, of the point upon the line.

3. Length being considered, the join of the projection of two points is the projection of the join of the points.

Thus if L be a given line and P, Q, two given points, and PP′, QQ′ perpendiculars upon L; PQ is the join of P and Q, P′ and Q′ are the projections of P and Q upon L, and the segment P′Q′ is the projection of PQ upon L.

168°. *Theorem.*—The sum of the projections of the sides of

any closed rectilinear figure, taken in cyclic order with respect to any line, is zero.

ABCD is a closed rectilinear figure and L is any line. Then
Pr.AB + Pr.BC + Pr.CD + Pr.DA = 0

Proof.—Draw the perpendiculars AA', BB', CC', DD', and the sum of the projections becomes
A'B' + B'C' + C'D' + D'A'.
But D'A' is equal in length to the sum of the three others and is opposite in sense. ∴ the sum is zero.

It is readily seen that since we return in every case to the point from which we start the theorem is true whatever be the number or disposition of the sides.

This theorem is of great importance in many investigations.

Cor. Any side of a closed rectilinear figure is equal to the sum of the projections of the remaining sides, taken in cyclic order, upon the line of that side.

Def.—In a right-angled triangle the side opposite the right angle is called the *hypothenuse*, as distinguished from the remaining two sides.

169°. *Theorem.*—In any right-angled triangle the square on one of the sides is equal to the rectangle on the hypothenuse and the projection of that side on the hypothenuse.

ABC is right-angled at B, and BD is ⊥ AC. Then $AB^2 = AC \cdot AD$.

Proof.—Let AF be the ☐ on AC, and let EH be ∥ to AB, and AGHB be a ▱, since ∠B is a ⌐.

Then ∵   ∠GAB = ∠EAC = ⌐,   (82°, Cor. 5)
∴   ∠CAB = ∠EAG.
Also,   AE = AC,   (hyp.)
∴   △CAB ≡ △EAG,   (64°)
and ∴   AG = AB, and AH is the ☐ on AB.

AREAL RELATIONS.

Now □AH = ⌷ABLE = ⌷ADKE,   (140°)
i.e., $AB^2 = AC \cdot AD$.   q.e.d.

As this theorem is very important we give an alternative proof of it.

Proof.—AF is the □ on AC and AH is the □ on AB, and BD is ⊥ AC.

∵ ∠GAB = ∠CAE = ⌐, (82°, Cor. 5)
∴ ∠GAC = ∠BAE.
Also, AG = AB,
and   AC = AE, (82°, Cor. 5)
∴ △GAC ≡ △BAE. (52°)
But △GAC = ½□AH, (141°)
and △BAE = ½⌷AK,
∴ □AH = ⌷AK,
i.e., $AB^2 = AC \cdot AD$.

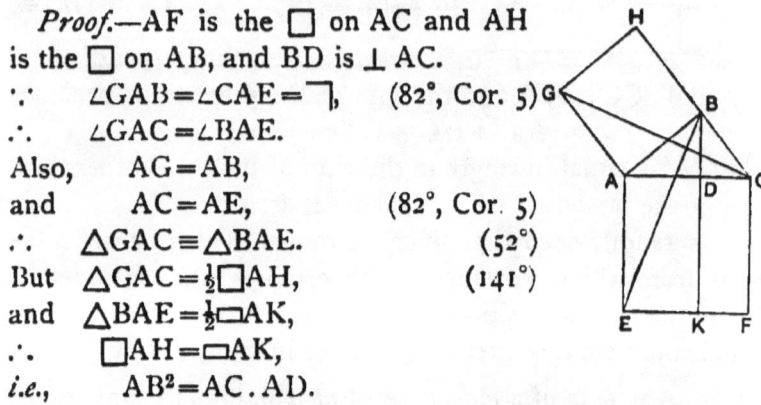

Cor. 1. Since $AB^2 = AC \cdot AD$ we have from symmetry
$$BC^2 = AC \cdot DC,$$
∴ adding, $AB^2 + BC^2 = AC(AD + DC),$
or $AB^2 + BC^2 = AC^2.$

∴ *The square on the hypothenuse of a right-angled triangle is equal to the sum of the squares on the remaining sides.*

This theorem, which is one of the most important in the whole of Geometry, is said to have been discovered by Pythagoras about 540 B.C.

Cor. 2. Denote the sides by a and c and the hypothenuse by b, and let a_1 and c_1 denote the projections of the sides a and c upon the hypothenuse.

Then $a^2 = a_1 b, \quad c^2 = c_1 b,$
and $a^2 + c^2 = b^2.$

Cor. 3. Denote the altitude to the hypothenuse by p.

Then $b = c_1 + a_1$, and ADB and CDB are right-angled at D,
∴ $b^2 = c_1^2 + a_1^2 + 2c_1 a_1;$ (166°, 1)
add $2p^2$ to each side and
$$b^2 + 2p^2 = c_1^2 + p^2 + a_1^2 + p^2 + 2c_1 a_1,$$

or $\quad c^2+a^2+2p^2=c^2+a^2+2c_1a_1.$ \hfill (Cor. 1)
∴ $\quad p^2=c_1a_1,$
or $\quad BD^2 = AD \cdot DC,$

i.e., the square on the altitude to the hypothenuse is equal to the rectangle on the projections of the sides on the hypothenuse.

Def.—The side of the square equal in area to a given rectangle is called the *mean proportional* or the *geometric mean* between the sides of the rectangle.

Thus the altitude to the hypothenuse of a right-angled △ is a geometric mean between the segments into which the altitude divides the hypothenuse. (169°, Cor. 3)

And any side of the △ is a geometric mean between the hypothenuse and its projection on the hypothenuse. (169°)

170°. *Theorem.*—If the square on one side of a triangle is equal to the sum of the squares on the remaining sides, the triangle is right-angled at that vertex which is opposite the side having the greatest square. (Converse of 169°, Cor.)

If $AC^2 = AB^2 + BC^2,$ the ∠B is a ⌐.

Proof.—Let ADC be a ½⊙ on AC.

∵ $\quad AC^2 = AB^2 + BC^2,$
$\quad AB$ is $< AC.$
∴ a chord AD can be found equal to AB.
Then the △ADC is right-angled at D. (106°, Cor. 4)

∴ $\quad AC^2 = AD^2 + DC^2,$ \hfill (169°, Cor. 1)
and $\quad AC^2 = AB^2 + BC^2,$ and $AD = AB.$ \hfill (hyp.)
∴ $\quad DC = BC,$
and $\quad \triangle ADC \equiv \triangle ABC.$
∴ $\quad \angle D = \angle B = \,⌐.$ \hfill *q.e.d.*

171°. Theorem 169° with its corollaries and theorem 170° are extensively employed in the practical applications of Geometry. If we take the three numbers 3, 4, and 5, we have $5^2 = 3^2 + 4^2$. Therefore if a triangle has its sides 3, 4, and 5 feet, metres, miles, or any other *u.l.*, it is right-angled opposite the side 5.

For the segments into which the altitude divides the hypothenuse we have $5a_1 = 3^2$ and $5c_1 = 4^2$, whence $a_1 = \frac{9}{5}$ and $c_1 = \frac{16}{5}$. For the altitude itself, $p^2 = \frac{9}{5} \cdot \frac{16}{5}$; whence $p = \frac{12}{5}$.

Problem.—To find sets of whole numbers which represent the sides of right-angled triangles.

This problem is solved by any three numbers x, y, and z, which satisfy the condition $x^2 = y^2 + z^2$.

Let m and n denote any two numbers. Then, since
$$(m^2 + n^2)^2 = (m^2 - n^2)^2 + (2mn)^2, \qquad (166°, 4)$$
the problem will be satisfied by the numbers denoted by $m^2 + n^2$, $m^2 - n^2$, and $2mn$.

The accompanying table, which may be extended at pleasure, gives a number of sets of such numbers:—

		\multicolumn{9}{c}{m}
		2	3	4	5	6	7	8	9	10	...
n	1	5 3 4	10 6 8	15 8 17	26 10 24	37 12 35	50 14 48	65 16 63	82 18 80	101 20 99	... ...
	2		13 12 5	20 16 12	29 20 21	40 24 32	53 28 45	68 32 60	85 36 77	104 40 96	... ...
	3			25 24 7	34 30 16	45 36 27	58 42 40	73 48 55	90 54 72	109 60 91	... ...
	4				41 9 40	52 48 20	65 56 33	80 64 48	97 72 65	116 80 84	... ...
	5					61 60 11	74 70 24	89 80 39	106 90 56	125 100 75	... ...
						...	...	...	...	...	...

172°. Let a, b, c be the sides of any triangle, and let b be taken as base. Denote the projections of a and c on b by a_1 and c_1, and the altitude to b by p. Then

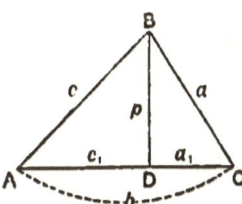

(1) $b^2 = c_1^2 + a_1^2 + 2c_1a_1$, (166°, 1)
(2) $c^2 = c_1^2 + p^2$, (169°, Cor. 1)
(3) $a^2 = a_1^2 + p^2$.

1. By subtracting (2) from (3)
$$a^2 - c^2 = a_1^2 - c_1^2.$$

∴ *The difference between the squares upon two sides of a triangle is equal to the difference of the squares on the projections of these sides on the third side, taken in the same order.*

Since all the terms are squares and cannot change sign (163°), the theorem is true without any variation for all △s.

2. By adding (1) and (2) and subtracting (3),
$$b^2 + c^2 - a^2 = 2c_1^2 + 2c_1a_1$$
$$= 2bc_1, \because b = c_1 + a_1,$$
∴ $a^2 = b^2 + c^2 - 2bc_1.$

Now, since we have assumed that $b = c_1 + a_1$, where c_1 and a_1 are both positive, D falls between A and C, and the angle A is acute.

∴ *In any triangle the square on a side opposite an acute angle is less than the sum of the squares upon the other two sides by twice the rectangle on one of these sides and the projection of the other side upon it.*

3. Let the angle A become obtuse. Then D, the foot of the altitude to b, passes beyond A, and c_1 changes sign.

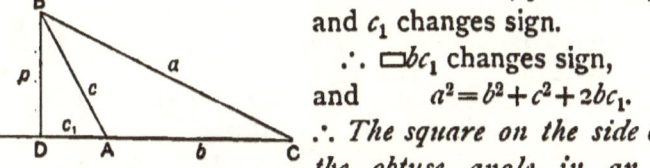

∴ $\Box bc_1$ changes sign, (161°)
and $a^2 = b^2 + c^2 + 2bc_1.$

∴ *The square on the side opposite the obtuse angle in an obtuse-angled triangle is greater than the sum of the squares on the other two sides by twice the rectangle on one of these sides and the projection of the other side upon it.*

The results of 2 and 3 are fundamental in the theory of triangles.

These results are but one; for, assuming as we have done that the $\square bc_1$ is to be subtracted from b^2+c^2 when A is an acute angle, the change in sign follows necessarily when A becomes obtuse, since in that case the $\square$ changes sign because one of its sides changes sign (161°); and in conformity to algebraic forms $-(-2bc_1) = +2bc_1$.

Cor. If the sides a, b, c of a triangle be given in numbers, we have from 2
$$c_1 = \frac{b^2+c^2-a^2}{2b},$$
which gives the projection of c on b.

If c_1 is $+$ the $\angle A$ is acute;
if c_1 is 0 the $\angle A$ is $\sqsupset$;
and if c_1 is $-$ the $\angle A$ is obtuse.

Ex. The sides of a triangle being 12, 13, and 4, to find the character of the angle opposite side 13.

Let $13 = a$, and denote the other sides as you please, *e.g.*, $b = 12$ and $c = 4$. Then
$$c_1 = \frac{12^2+4^2-13^2}{24} = -\frac{3}{8},$$
and the angle opposite side 13 is obtuse.

173°. *Theorem.* —The sum of the squares on any two sides of a triangle is equal to twice the sum of the squares on one-half the third side and on the median to that side.

BE is the median to AC. Then
$$AB^2 + BC^2 = 2(AE^2 + EB^2).$$

Proof.—Let D be the foot of the altitude on AC. Consider the $\triangle ABE$ obtuse-angled at E, and
$$AB^2 = AE^2 + EB^2 + 2AE \cdot ED. \qquad (172°, 3)$$
Next, consider the $\triangle CBE$ acute-angled at E, and
$$BC^2 = EC^2 + EB^2 - 2EC \cdot ED. \qquad (172°, 2)$$

Now, adding and remembering that AE = EC,
$$AB^2 + BC^2 = 2AE^2 + 2EB^2.$$ q.e.d.

Cor. 1. Denoting the median by m and the side upon which it falls by b, we have for the length of the median
$$m^2 = \frac{2(a^2 + c^2) - b^2}{4}.$$

Cor. 2. All the sides of an equilateral triangle are equal and the median is the altitude to the base and the right bisector of the base. (53°, Cors. 2, 3)

∴ in an equilateral triangle,
$$m^2 = p^2 = \tfrac{3}{4}a^2, \text{ or } p = \tfrac{1}{2}a\sqrt{3}, \ a \text{ being the side.}$$

173°. Theorem.—The sum of the squares on the sides of a quadrangle is equal to the sum of the squares on the diagonals, and four times the square on the join of the middle points of the diagonals.

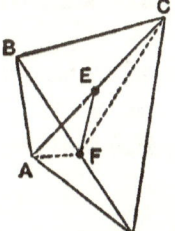

E, F are middle points of AC and BD.
Then $\Sigma(AB^2) = AC^2 + BD^2 + 4EF^2$.

Proof.—Join AF and CF.
Then AF is a median to $\triangle ABD$, and CF to $\triangle CBD$.
∴ $\quad AB^2 + AD^2 = 2BF^2 + 2AF^2,\quad$ (172°)
and $\quad BC^2 + CD^2 = 2BF^2 + 2CF^2,$
∴ adding, $\quad \Sigma(AB^2) = 4BF^2 + 2(AF^2 + CF^2).$
But EF is a median to $\triangle AFC$.
∴ $\quad AF^2 + CF^2 = 2CE^2 + 2EF^2,\quad$ (172°)
∴ $\quad \Sigma(AB^2) = 4BF^2 + 4CE^2 + 4EF^2$
$\quad\quad\quad\quad = BD^2 + AC^2 + 4EF^2.\quad$ q.e.d.

Since squares only are involved this relation is true without any modification for all quadrangles.

Cor. 1. When the quadrangle becomes a ▭ the diagonals bisect one another (81°, 3) and EF becomes zero.

∴ the sum of the squares on the sides of a parallelogram is equal to the sum of the squares on its diagonals.

174°. Let ABC be an isosceles triangle and P be any point in the base AC, and let D be the middle point of the base, and therefore the foot of the altitude.

In the $\triangle$BAP, acute-angled at A,
$$BP^2 = BA^2 + AP^2 - 2AP \cdot AD, \quad (172°, 2)$$
$\therefore \quad BA^2 - BP^2 = AP(2AD - AP)$
$$= AP \cdot PC.$$

If P moves to Q, AP becomes AQ and changes sign, BP becomes BQ which is > BA, and thus both sides of the equality change sign together as they pass through zero by P passing A.

Now, of the two segments from B we always know which is the greater by 63°, and if we write PA for AP the $\square$PA.PC is positive when P is on the Q side of A. Hence, considering the rectangle as being always positive, we may state the theorem—

The difference between the squares on a side of an isosceles triangle and on the join of the vertex to any point in the base is equal to the rectangle on the segments into which that point divides the base.

175°. 1. From 174° we have $BA^2 - BP^2 = AP \cdot PC$. Now BA is fixed, therefore the $\square$AP.PC increases as BP decreases. But BP is least when P is at D (63°, 1), therefore the $\square$AP.PC is greatest when P is at D.

Def. 1.—A variable magnitude, which by continuous change may increase until a greatest value is reached and then decrease, is said to be capable of a maximum, and the greatest value reached is its *maximum*.

Thus as P moves from A to C the $\square$AP.PC increases from zero, when P is at A, to its maximum value, when P is at D, and then decreases again to zero, when P comes to C.

And as AC may be considered to be any segment divided at P,

$\therefore$ *The maximum rectangle on the parts of a given segment is formed by bisecting the segment;*

Or, *of all rectangles with a given perimeter the square has the greatest area.*

2. $\quad AC^2 = (AP+PC)^2 = AP^2+PC^2+2AP \cdot PC \quad$ (166°, 1)
$\quad\quad\quad = AP^2+PC^2+2(AB^2-BP^2).\quad$ (174°)

But AC and AB are constant,

∴ $\quad AP^2+PC^2$ decreases as BP^2 decreases.

But BP is least when P is at D,

∴ $\quad AP^2+PC^2$ is least when P is at D.

Def. 2.—A variable magnitude which by continuous change decreases until it reaches a least value and then increases is said to be *capable of a minimum*, and the least value attained is called its *minimum*.

∴ *The sum of the squares on the two parts of a given segment is a minimum when the segment is bisected.*

175½°. The following examples give theorems of importance.

Ex. 1. Let ABC be any triangle and BD the altitude to side b. Then
$$c_1 = \frac{b^2+c^2-a^2}{2b}, \quad \text{(172°, Cor.)}$$

But $\quad p^2 = c^2 - c_1^2 = (c+c_1)(c-c_1),$

and $\quad \triangle = \tfrac{1}{2}bp.\quad$ (153°, 2)

Now, $\quad c+c_1 = \dfrac{(b+c)^2-a^2}{2b} = \dfrac{(b+c+a)(b+c-a)}{2b},$

and $\quad c-c_1 = \dfrac{a^2-(b-c)^2}{2b} = \dfrac{(a+b-c)(a-b+c)}{2b}.$

∴ $\quad 4b^2p^2 = 16\triangle^2 = (a+b+c)(b+c-a)(c+a-b)(a+b-c),$

and by writing s for $\tfrac{1}{2}(a+b+c)$, and accordingly $s-a$ for $\tfrac{1}{2}(b+c-a)$, etc., we obtain
$$\triangle = \sqrt{s(s-a)(s-b)(s-c)}.$$

This important relation gives the area of the $\triangle$ in terms of its three sides.

Ex. 2. Let ABC be an equilateral $\triangle$. Then the area may be found from Ex. 1 by making $a=b=c$, when the reduced expression becomes, $\quad \triangle = \dfrac{a^2}{4}\sqrt{3}.$

AREAL RELATIONS.

Ex. 3. To find the area of a regular octagon in terms of its circumradius.

Let A, B, C be three vertices of the octagon and O the centre. Complete the square OD, and draw BE $\perp$ to OA.

Since $\angle EOB = \frac{1}{2}\llcorner$, and $OB = r$,

$\therefore \qquad EO = EB = \frac{1}{2}r\sqrt{2}$,

and $\triangle OAB = \frac{1}{2}OA \cdot EB = \frac{1}{2}r \cdot \frac{1}{2}r\sqrt{2} = \frac{1}{4}r^2\sqrt{2}$.

But $\triangle OAB$ is one-eighth of the octagon,

$\therefore \qquad$ Oct. $= 2r^2\sqrt{2}$.

EXERCISES.

1. ABC is right-angled at B, and E and F are middle points of BA and BC respectively. Then $5AC^2 = 4(CE^2 + AF^2)$.
2. ABC is right-angled at B and O is the middle of AC, and D is the foot of the altitude from B. Then $2AC \cdot OD = AB^2 - BC^2$.
3. ABC is right-angled at B and, on AC, AD is taken equal to AB, and on CA, CE is taken equal to CB. Then $ED^2 = 2AE \cdot DC$.
4. The square on the sum of the sides of a right-angled triangle exceeds the square on the hypotenuse by twice the area of the triangle.
5. To find the side of a square which is equal to the sum of two given squares.
6. To find the side of a square which is equal to the difference of two given squares.
7. The equilateral triangle described upon the hypothenuse of a right-angled triangle is equal to the sum of the equilateral triangles described on the sides.
8. ABC is a triangle having AB = CB, and AD is $\perp$ upon BC. Then $AC^2 = 2CB \cdot CD$.
9. Four times the sum of the squares on the three medians of a triangle is equal to three times the sum of the squares on the sides.

10. ABCD is a rectangle and P is any point. Then
$$PA^2 + PC^2 = PB^2 + PD^2.$$
11. O is the centre of a circle, and AOB is a centre-line. $OA = OB$ and C is any point on the circle. Then $AC^2 + BC^2 =$ a constant.

   Define a circle as the locus of the point C.
12. AD is a perpendicular upon the line OB, and BE is a perpendicular upon the line OA. Then $OA \cdot OE = OB \cdot OD$.
13. Two equal circles pass each through the centre of the other. If A, B be the centres and E, F be the points of intersection, $EF^2 = 3AB^2$.

   If EA produced meets one circle in P and AB produced meets the other in Q, $PQ^2 = 7AB^2$.
14. ABC is a triangle having the angle A two-thirds of a right angle. Then $AB^2 + AC^2 = BC^2 + AC \cdot AB$.
15. In the triangle ABC, D is the foot of the altitude to AC and E is the middle point of the same side. Then $2ED \cdot AC = AB^2 - BC^2$.
16. AD is a line to the base of the triangle ABC, and O is the middle point of AD. If $AB^2 + BD^2 = AC^2 + CD^2$, then $OB = OC$.
17. ABC is right-angled at B and BD is the altitude to AC. Then $AB \cdot CD = BD \cdot BC$ and $AD \cdot CB = BA \cdot BD$.
18. ABC is a triangle and OX, OY, OZ perpendiculars from any point O on BC, CA, and AB respectively. Then
$$BX^2 + CY^2 + AZ^2 = CX^2 + AY^2 + BZ^2.$$

   A similar relation holds for any polygon.
19. AA_1, BB_1 are the diagonals of a rectangle and P any point. Then $PA^2 + PB^2 + PA_1^2 + PB_1^2 = AA_1^2 + 4PO^2$, where O is the intersection of the diagonals.
20. ABC is a triangle, AD, BE, CF its medians, and P any point. Then
$$PA^2 + PB^2 + PC^2 = PD^2 + PE^2 + PF^2 + \tfrac{1}{3}(AD^2 + BE^2 + CF^2),$$
   or $\Sigma PA^2 = \Sigma PD^2 + \tfrac{1}{3}\Sigma m^2$, where m is a median.
21. If O be the centroid in 20,

$$PA^2 + PB^2 + PC^2 = 3PO^2 + \tfrac{4}{9}(AD^2 + BE^2 + CF^2),$$
or
$$\Sigma PA^2 = 3PO^2 + \tfrac{4}{9}\Sigma m^2.$$

22. ABCD is a square and AA', BB', CC', DD' perpendiculars upon any line L. Then
$$(AA'^2 + CC'^2) \sim 2BB'.DD' = \text{area of the square}.$$

23. The sum of the squares on the diagonals of any quadrangle is equal to twice the sum of the squares on the joins of the middle points of opposite sides.

24. ABCD is a trapezoid having AD parallel to BC. Then
$$AB^2 + CD^2 + 2AD.BC = AC^2 + BD^2.$$

25. If A, B, C be equidistant points in line, and D a fourth point in same line, the difference between the squares on AB and DB is equal to the rectangle on AD and CD.

26. If A, B, C, D be any four points in line,
$$AD^2 + BC^2 = AC^2 + BD^2 + 2AB.CD.$$

27. Any rectangle is equal to one-half the rectangle on the diagonals of the squares described on adjacent sides.

28. In the triangle ABC, D is any point in BC, E is the middle point of AC and F of BC. Then
$$AB^2 + AC^2 = AD^2 + 4EF^2 + 2BD.DC.$$

29. The sides of a rectangle are a and b. If p be the length of the perpendicular from a vertex upon a diagonal and q be the distance between the feet of the two parallel perpendiculars so drawn,
$$p\sqrt{a^2+b^2} = ab \text{ and } q\sqrt{a^2+b^2} = b^2 - a^2 \ (b > a),$$
what line-segment is denoted by $\sqrt{a^2+b^2}$?

30. ABCD is a square. P is a point in AB produced, and Q is a point in AD. If the rectangle BP.QD is constant, the triangle PQC is constant.

31. If the lengths of the sides of a triangle be expressed by x^2+1, x^2-1, and $2x$, the triangle is right-angled.

32. If a and c be the sides of a right-angled triangle and p be the altitude to the hypothenuse,
$$\frac{1}{a^2} + \frac{1}{c^2} = \frac{1}{p^2}.$$

33. The triangle whose sides are 20, 15, and 12 has an obtuse angle.
34. The area of an isosceles triangle is $8\sqrt{15}$ and the side is twice as long as the base. Find the length of the side of the triangle.
35. What is the length of the side of an equilateral triangle which is equal to the triangle whose sides are 13, 14, and 15?
36. If AB is divided in C so that $AC^2 = 2BC^2$, then
$$AB^2 + BC^2 = 2AB \cdot AC.$$
37. Applying the principle of continuity state the resulting theorem when B comes to D in (1) the Fig. of 172°, (2) the Fig. of 173°.
38. Applying the principle of continuity state the resulting theorem when B comes to E in the Fig. of 173°.
39. The bisector of the right angle of a right-angled triangle cuts the hypothenuse at a distance a from the middle point, and the hypothenuse is $2b$. Find the lengths of the sides of the triangle.
40. Construct an equilateral triangle having one vertex at a given point and the remaining vertices upon two given parallel lines.
41. A square of cardboard whose side is s stands upright with one edge resting upon a table. If a lower corner be raised vertically through a distance a, through what distance will the corner directly above it be raised?
42. What would be the expression for the area of a rectangle if the area of the equilateral triangle having its side the *u.l.* were taken as the *u.a.*?
43. The opposite walls of a house are 12 and 16 feet high and 20 feet apart. The roof is right-angled at the ridge and has the same inclination on each side. Find the lengths of the rafters.
44. Two circles intersect in P and Q. The longest chord through P is perpendicular to PQ.

AREAL RELATIONS.

45. The largest triangle with a given perimeter is an equilateral triangle.
46. The largest triangle having its base and the sum of the other two sides given is isosceles.
47. The largest polygon of given species and given perimeter is regular.
48. The largest isosceles triangle with variable base has its sides perpendicular to one another.
49. The largest rectangle inscribed in an acute-angled triangle and having one side lying on a side of the triangle has its altitude one-half that of the triangle.
50. L, M are two lines meeting in O, and P is any point. APB is a variable line cutting L in A and M in B. The triangle AOB is least when P bisects AB.

EQUALITIES OF RECTANGLES ON SEGMENTS RELATED TO THE CIRCLE.

176°. *Theorem.*—If two secants to the same circle intersect, the rectangle on the segments between the point of intersection and the circle with respect to one of the secants is equal to the corresponding rectangle with respect to the other secant.

1. Let the point of intersection be within the circle. Then
$$AP \cdot PB = CP \cdot PD.$$

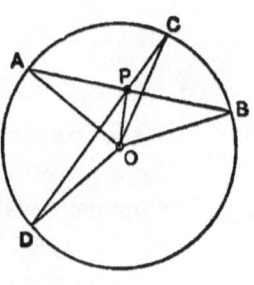

Proof.—AOB is an isosceles triangle, and P is a point on the base AB.

∴ $OA^2 - OP^2 = AP \cdot PB.$ (174°)

Similarly, COD is an isosceles triangle, and P a point in the base CD,

∴ $OC^2 - OP^2 = CP \cdot PD.$
But $OC = OA,$
∴ $AP \cdot PB = CP \cdot PD.$ *q.e.d.*

Cor. 1. (*a*) Let CD become a diameter and be ⊥ to AB.

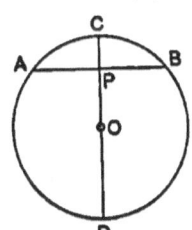

Then AP . PB becomes AP^2, (96°, Cor. 5)
∴ $AP^2 = CP \cdot PD$,
and denoting AP by c, CP by v, and the radius of the circle by r, this becomes
$$c^2 = v(2r - v),$$
which is a relation between a chord of a ⊙, the radius of the ⊙, and the distance CP, commonly called the *versed sine*, of the arc AB.

(*b*) When the point of intersection P passes without the ⊙

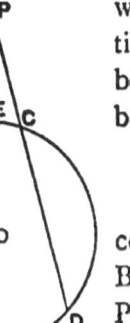

we have still, by the principle of continuity, AP . PB = CP . PD. But the ▭s being now both negative we make them both positive by writing
$$PA \cdot PB = PC \cdot PD.$$

Cor. 2. When the secant PAB becomes the tangent PT (109°), A and B coincide at T, and PA . PB becomes PT^2, ∴ $PT^2 = PC \cdot PD$,

i.e., *if a tangent and a secant be drawn from the same point to a circle, the square on the tangent is equal to the rectangle on the segments of the secant between the point and the circle.*

Cor. 3. Conversely, if T is on the circle and $PT^2 = PC \cdot PD$, PT is a tangent and T is the point of contact.

For, if the line PT is not a tangent it must cut the circle in some second point T′ (94°). Then
$$PT \cdot PT' = PC \cdot PD = PT^2.$$
Therefore PT = PT′, which is not true unless T and T′ coincide. Hence PT is a tangent and T is the point of contact.

Cor. 4. Let one of the secants become a centre-line as PEF. Denote PT by t, PE by h, and the radius of the circle by r. Then $PT^2 = PE \cdot PF$
becomes $t^2 = h(2r + h)$.

AREAL RELATIONS.

EXERCISES.

1. The shortest segment from a point to a circle is a portion of the centre-line through the point.
2. The longest segment from a point to a circle is a portion of the centre-line through the point.
3. If two chords of a circle are perpendicular to one another the sum of the squares on the segments between the point of intersection and the circle is equal to the square on the diameter.
4. The span of a circular arch is 120 feet and it rises 15 feet in the middle. With what radius is it constructed?
5. A conical glass is b inches deep and a inches across the mouth. A sphere of radius r is dropped into it. How far is the centre of the sphere from the bottom of the glass?
6. The earth's diameter being assumed at 7,960 miles, how far over its surface can a person see from the top of a mountain 3 miles high?
7. How much does the surface of still water fall away from the level in one mile?
8. Two circles whose radii are 10 and 6 have their centres 12 feet apart. Find the length of their common chord, and also that of their common tangent.
9. Two parallel chords of a circle are c and c_1 and their distance apart is d, to find the radius of the circle.
10. If v is the versed sine of an arc, k the chord of half the arc, and r the radius, $k^2 = 2vr$.

177°. *Theorem.*—If upon each of two intersecting lines a pair of points be taken such that the rectangle on the segments between the points of intersection and the assumed points in one of the lines is equal to the corresponding rectangle for the other line, the four assumed points are concyclic.

(Converse of 176°.)

L and M intersect in O, and
OA . OB = OC . OD.
Then A, B, C, and D are concyclic.

Proof.—Since the ▭s are equal, if A and B lie upon the same side of O, C and D must lie upon the same side of O; and if A and B lie upon opposite sides of O, C and D must lie upon opposite sides of O.

Let a ⊙ pass through A, B, C, and let it cut M in a second point E.   Then   OA . OB = OC . OE.   (176°)
But   OA . OB = OC . OD.   (hyp.)
∴   OD = OE,

and as D and E are upon the same side of O they must coincide;   ∴ A, B, C, D are concyclic.   *q.e.d.*

178°. Let two circles excluding each other without contact have their centres at A and B, and let C be the point, on their common centre-line, which divides AB so that the difference between the squares on the segments AC and CB is equal to the difference between the squares on the conterminous radii. Through C draw the line PCD ⊥ to AB, and from any point P on this line draw tangents PT and PT' to the circles.

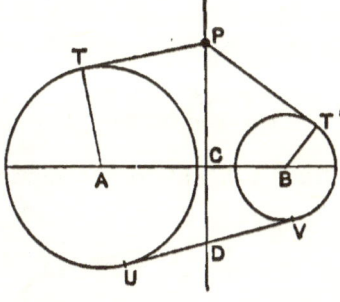

Join AT and BT'.
Then, by construction,
$AC^2 - BC^2 = AT^2 - BT'^2$.

But, since PC is an altitude in the △APB, $AC^2 - BC^2 = AP^2 - BP^2$,   (172°, 1)
and $AP^2 = AT^2 + PT^2$,
and $BP^2 = BT'^2 + PT'^2$,   (169°, Cor. 1)
whence $PT^2 = PT'^2$,
and   PT = PT'.

AREAL RELATIONS. 137

Therefore PCD is the locus of a point from which equal tangents are drawn to the two circles.

Def.—This locus is called the *radical axis* of the circles, and is a line of great importance in studying the relations of two or more circles.

Cor. 1. The radical axis of two circles bisects their common tangents.

Cor. 2. When two circles intersect, their radical axis is their common chord.

Cor. 3. When two circles touch externally, the common tangent at the point of contact bisects the other common tangents.

179°. The following examples give theorems of some importance.

Ex. 1. P is any point without a circle and TT' is the chord of contact (114°, Def.) for the point P. TT' cuts the centre-line PO in Q. Then, PTO being a ⌐, (110°)
$$OQ \cdot OP = OT^2. \quad (169°)$$
∴ *the radius is a geometric mean between the join of any point with the centre and the perpendicular from the centre upon the chord of contact of the point.*

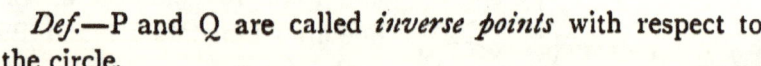

Def.—P and Q are called *inverse points* with respect to the circle.

Ex. 2. Let PQ be a common direct tangent to the circles having O and O' as centres.

Let OP and O'Q be radii to the points of contact, and let QR be ∥ to OO'. Denote the radii by r and r'. Then
$$AC = OO' + r - r',$$
$$BD = OO' - r + r'.$$

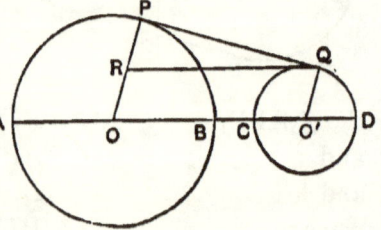

∴ $AC \cdot BD = OO'^2 - (r-r')^2 = QR^2 - PR^2 = PQ^2.$ (169°, Cor. 1)

Similarly it may be shown that

AD . BC = square on the transverse common tangent.

Exercises.

1. The greater of two chords in a circle is nearer the centre than the other.
2. Of two chords unequally distant from the centre the one nearer the centre is the greater.
3. AB is the diameter of a circle, and P, Q any two points on the curve. AP and BQ intersect in C, and AQ and BP in C'. Then
$$AP . AC + BQ . BC = AC' . AQ + BC' . BP.$$
4. Two chords of a circle, AB and CD, intersect in O and are perpendicular to one another. If R denotes the radius of the circle and E its centre,
$$8R^2 = AB^2 + CD^2 + 4OE^2.$$
5. Circles are described on the four sides of a quadrangle as diameters. The common chord of any two adjacent circles is parallel to the common chord of the other two.
6. A circle S and a line L, without one another, are touched by a variable circle Z. The chord of contact of Z passes through that point of S which is farthest distant from L.
7. ABC is an equilateral triangle and P is any point on its circumcircle. Then PA + PB + PC = o, if we consider the line crossing the triangle as being negative.
8. CD is a chord parallel to the diameter AB, and P is any point in that diameter. Then
$$PC^2 + PD^2 = PA^2 + PB^2.$$

SECTION V.

CONSTRUCTIVE GEOMETRY.

180°. Problem.—AB being a given segment, to construct the segment AB√2.

Constr.—Draw BC ⊥ to AB and equal to it. Then AC is the segment AB√2.

Proof.—Since ABC is right-angled at B,
$$AC^2 = AB^2 + BC^2 = 2AB^2, \quad (169, \text{ Cor. } 1)$$
∴ $AC = AB\sqrt{2}$.

Cor. The square on the diagonal of a given square is equal to twice the given square.

181°. Problem.—To construct AB√3.

Constr.—Take BC in line with AB and equal to it, and on AC construct an equilateral triangle ADC. (124°, Cor. 1)

BD is the segment AB√3.

Proof.—ABD is a ⌐, and AD = AC = 2AB.
Also $AD^2 = AB^2 + BD^2 = 4AB^2$. (169°, Cor. 1)
∴ $BD^2 = 3AB^2$, and $BD = AB\sqrt{3}$.

Cor. Since BD is the altitude of an equilateral triangle and AB is one-half the side,

∴ the square on the altitude of an equilateral triangle is equal to three times the square on the half side.

182°. Problem.—To construct AB√5.

Constr.—Draw BC ⊥ to AB and equal to twice AB. Then AC is the segment AB√5.

Proof.—Since ∠B is a right angle,
$$AC^2 = AB^2 + BC^2.$$
But $BC^2 = 4AB^2$;
∴ $AC^2 = 5AB^2$,
and $AC = AB\sqrt{5}$.

183°. The three foregoing problems furnish *elements* of construction which are often convenient. A few examples are given.

Ex. 1. AB being a given segment, to find a point C in its line such that $AC^2 = AB \cdot CB$.

Analysis— $AC^2 = AB \cdot CB = AB(AB - AC)$,
∴ $AC^2 + AC \cdot AB = AB^2$.

Considering this an algebraic form and solving as a quadratic in AC, we have $AC = \tfrac{1}{2}(AB\sqrt{5} - AB)$, and this is to be constructed.

Constr.—Construct $AD = AB\sqrt{5}$ (by 182°) as in the figure, and let E be the middle point of BD.

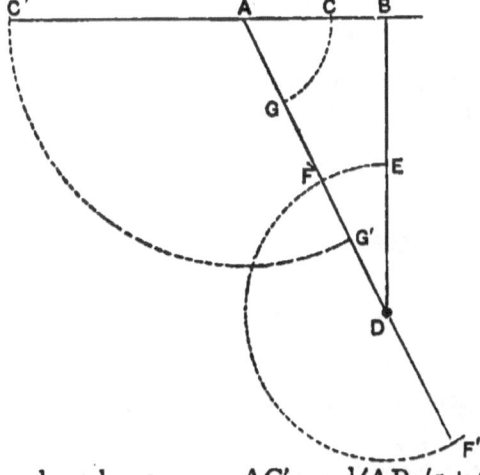

Take $DF = DE$. Then
$AF = AB\sqrt{5} - AB$;
∴ bisecting AF in G,
$AG = AC$
$= \tfrac{1}{2}(AB\sqrt{5} - AB)$,
and the point C is found.

Again, since $\sqrt{5}$ has two signs + or −, take its negative sign and we have $AC' = -\tfrac{1}{2}(AB\sqrt{5} + AB)$.

Therefore, for the point C′, on AD produced take $DF' = DE$, and bisect AF′ in G′. Then
$AG' = \tfrac{1}{2}(AB\sqrt{5} + AB)$;
and since AC′ is negative we set off AG′ from A to C′, and C′ is a second point.

The points C and C′ satisfy the conditions,
$AC^2 = AB \cdot CB$ and $AC'^2 = AB \cdot C'B$.

A construction effected in this way requires no proof other than the equation which it represents.

It is readily proved however. For
$AD^2 = 5AB^2$, and also $AD^2 = (AF+FD)^2 = (2AC+AB)^2$,
whence $AC^2 = AB(AB-AC) = AB \cdot CB$.

It will be noticed that the constructions for finding the two points differ only by some of the segments being taken in different senses. Thus, for C, DE is taken from DA, and for C′, added to DA; and for C, AC is taken in a positive sense equal to AG, and for C′, AC′ is taken in a negative sense equal to AG′.

In connection with the present example we remark :—

1. Where the analysis of a problem involves the solution of a quadratic equation, the problem has two solutions corresponding to the roots of the equation.

2. Both of the solutions may be applicable to the wording of the problem or only one may be.

3. The cause of the inapplicability of one of the solutions is commonly due to the fact that a mathematical symbol is more general in its significance than the words of a spoken language.

4. Both solutions may usually be made applicable by some change in the wording of the problem so as to generalize it.

The preceding problem may be stated as follows, but whether both solutions apply to it, or only one, will depend upon our definition of the word "part." See Art. 23°.

To divide a given segment so that the square upon one of the parts is equal to the rectangle on the whole segment and the other part.

Def.—A segment thus divided is said to be divided into *extreme and mean ratio*, or in *median section*.

Ex. 2. To describe a square when the sum of its side and diagonal is given.

Analysis.—If AB is the side of a square, $AB\sqrt{2}$ is its diagonal, (180°)

∴ AB($1 + \sqrt{2}$) is a given segment = S, say. Then
AB = S($\sqrt{2} - 1$).

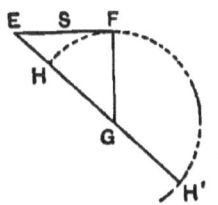

Constr.—Let EF be the given segment S. Draw FG ⊥ and = to EF, and with centre G and radius GF describe a ⊙ cutting EG in H and H'.

EH is the side of the square; whence the square is easily constructed.

If we enquire what EH' means, we find it to be the side of the square in which the *difference* between the side and diagonal is the given segment S. The double solution here is very suggestive, but we leave its discussion to the reader.

184°. *Problem.*—To find a segment such that the rectangle on it and a given segment shall be equal to a given rectangle.

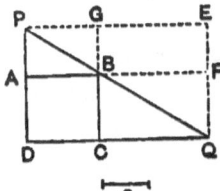

Constr.—Let S be the given segment, and AC the given rectangle.

On DA produced make AP = S, and draw PBQ to cut DC produced in Q.

CQ is the segment required.

Proof.—Complete the ▭s PEQD, PGBA, and BCQF.
Then ▭AC = ▭GF = GB . BF = PA . CQ,
∴ S . CQ = ▭AC.

Def.—The segments AP and CQ are *reciprocals* of one another with respect to the ▭AC as unit.

185°. *Problem.*—To find the side of a square which is equal to a given rectangle.

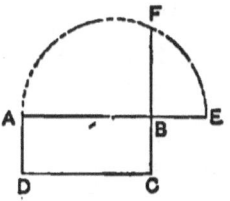

Constr.—Let AC be the rectangle. Make BE = BC and in line with BA. On AE describe a semicircle, and produce CB to meet it in F.

BF is the side of the required square.

Proof.—Since AE is a diameter and FB a half chord ⊥ to it,

CONSTRUCTIVE GEOMETRY.   143

$$BF^2 = AB \cdot BE, \quad (176°, \text{Cor. 1})$$
$$\therefore \quad BF^2 = AB \cdot BC.$$

Cor. This is identical with the problem, "To find a geometric mean between two given segments," and it furnishes the means of constructing the segment a, when $a = \sqrt{bc}$, b and c being given.   (165°)

Ex. 1. To construct an equilateral triangle equal to a given rectangle.

Let AC be the given rectangle, and suppose PQR to be the required triangle. Then

$$AB \cdot BC = \tfrac{1}{2} PR \cdot QT$$
$$= PT \cdot QT.$$

But $QT = PT\sqrt{3},$ (181°, Cor.)

$\therefore \quad PT \cdot QT = PT^2 \sqrt{3}$

whence $PT^2 = AB\sqrt{3} \cdot \tfrac{1}{3} BC.$

And PT is the side of a square equal to the rectangle whose sides are $AB\sqrt{3}$ and $\tfrac{1}{3}BC$, and is found by means of 181°, 127°, and 185°.

Thence the triangle is readily constructed.

Ex. 2. To bisect the area of a triangle by a line parallel to its base.

Let ABC be the triangle, and assume PQ as the required line, and complete the parallelograms AEBC, KFBC, and let BD be the altitude to AC. Because PQ is ∥ to AC, BD is ⊥ to PQ. Now

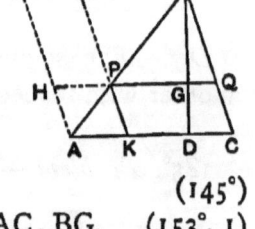

$$\square EP = \square PC, \quad (145°)$$
$$\therefore \quad \square FC = \square EQ, \text{ or } PQ \cdot BD = AC \cdot BG. \quad (153°, 1)$$

But $2\square FQ = \square EC,$ or $2PQ \cdot BG = AC \cdot BD$;

∴ dividing one equation by the other, and reducing to one line, $BD^2 = 2BG^2$;

and therefore BG is one-half the diagonal of the square of which BD is the side, and the position of PQ is determined.

144 SYNTHETIC GEOMETRY.

186°. *Problem.* —To find the circle which shall pass through two given points and touch a given line.

Let A, B be the given points and L the given line.

Constr.—Let the line AB cut L in O. Take OP=OP', a geometric mean between OA and OB (185°). The circles through the two sets of three points A, B, P and A, B, P' are the two solutions.

The proof is left to the reader. (See 176°, Cor. 2.)

187°. *Problem.* —To find a $\odot$ to pass through two given points and touch a given $\odot$.

Let A, B be the points and S the given $\odot$.

Constr.—Through A and B draw any $\odot$ so as to cut S in two points C and D. Let the line CD meet the line AB in O. From O draw tangents OP and OQ to the $\odot$S (114°). P and Q are the points of contact for the $\odot$s which pass through A and B and touch S. Therefore the $\odot$s through the two sets of three points A, B, P and A, B, Q are the $\odot$s required.

Proof.— OB.OA=OC.OD=OQ^2=OP^2;
therefore the $\odot$s through A, B, P and A, B, Q have OP and OQ as tangents (176°, Cor. 3). But these are also tangents to $\odot$S; therefore P and Q are the points of contact of the required $\odot$s.

EXERCISES.

1. Describe a square that shall have twice the area of a given square.
2. Describe an equilateral triangle equal to a given square.

3. Describe an equilateral triangle having five times the area of a given equilateral triangle.
4. Construct $AB\sqrt{7}$, where AB is a given segment.
5. Construct $\sqrt{a^2+b^2}$ and $\sqrt{a^2-b^2}$, where a and b denote given line segments.
6. Divide the segment AB in C so that $AC^2 = 2CB^2$. Show that AC is the diagonal of the square on CB. Does this hold for external division also?
7. ABCD is a rectangle and DE, a part of DA, is equal to DC. EF, perpendicular to AD, meets the circle having A as centre and AD as radius in F. Then DF is the diagonal of a square equal to the rectangle.
8. In the Fig. of 183°, $CE^2 = 3AB \cdot CB$,
   $CD^2 = CE^2 + 3ED^2 = 3AB(AB+CB)$.
9. Show that the construction of 183° solves the problem, "To divide a segment so that the rectangle on the parts is equal to the difference of the squares on the parts."
10. Show that the construction of 183° solves the problem, "To divide a given segment so that the rectangle on the whole and one of the parts is equal to the rectangle on the other part and the segment which is the sum of the whole and the first part."
11. Construct an equilateral triangle when the sum of its side and altitude are given. What does the double solution mean? (See 183°, Ex. 2.)
12. Describe a square in a given acute-angled triangle, so that one side of the square may coincide with a side of the triangle.
13. Within a given square to inscribe a square having three-fourths the area of the first.
14. Within an equilateral triangle to inscribe a second equilateral triangle whose area shall be one-half that of the first.
15. Produce a segment AB to C so that the rectangle on the sum and difference of AC and AB shall be equal to a given square.

K

16. Draw a tangent to a given circle so that the triangle formed by it and two fixed tangents may be (1) a maximum, (2) a minimum.
17. Draw a circle to touch two sides of a given square, and pass through one vertex. Generalize this problem and show that there are two solutions.
18. Given any two lines at right angles and a point, to find a circle to touch the lines and pass through the point.
19. Describe a circle to pass through a given point and to touch a given line at a given point in the line.
20. Draw the oblique lines required to change a given square into an octagon.

    If the side of a square is 24, the side of the resulting octagon is approximately 10; how near is the approximation?
21. The area of a regular dodecagon is three times that of the square on its circumradius.
22. By squeezing in opposite vertices of a square it is transformed into a rhombus of one-half the area of the square. What are the lengths of the diagonals of the rhombus?
23. P, Q, R, S are the middle points of the sides AB, BC, CD, and DA of a square. Compare the area of the square with that of the square formed by the joins AQ, BR, CS, and DP.
24. ABCDEFGH is a regular octagon, and AD and GE are produced to meet in K. Compare the area of the triangle DKE with that of the octagon.
25. The rectangle on the chord of an arc and the chord of its supplement is equal to the rectangle on the radius and the chord of twice the supplement.
26. At one vertex of a triangle a tangent is drawn to its circumcircle. Then the square on the altitude from that vertex is equal to the rectangle on the perpendiculars from the other vertices to the tangent.
27. SOT is a centre-line and AT a tangent to a circle at the point A. Determine the angle AOT so that AS=AT.

PART III.

PRELIMINARY.

188°. By superposition we ascertain the equality or inequality of two given line-segments. But in order to express the relation between the lengths of two unequal segments we endeavour to find two numerical quantities which hold to one another the same relations in magnitude that the given segments do.

Let AB and CD be two given segments. If they are commensurable (150°, 5) some *u.l.* can be found with respect to which the measures of AB and CD (150°, 2) are both whole numbers. Let *m* denote the measure of AB and *n* the measure of CD with respect to this unit-length.

The numbers *m* and *n* hold to one another the same relations as to magnitude that the segments AB and CD do.

The fraction $\frac{m}{n}$ is called in Arithmetic or Algebra the *ratio* of *m* to *n*, and in Geometry it is called the ratio of AB to CD.

Now *n* has to *m* the same ratio as unity has to the fraction $\frac{m}{n}$. But if CD be taken as *u.l.* its measure becomes unity, while that of AB becomes $\frac{m}{n}$.

Therefore the *ratio* of AB to CD is the measure of AB with respect to CD as unit-length.

When AB and CD are commensurable this ratio is expressible arithmetically either as a whole number or as a fraction;

but when the segments are incommensurable the ratio can only be symbolized, and cannot be expressed arithmetically except approximately.

189°. If we suppose CD to be capable of being stretched until it becomes equal in length to AB, the numerical factor which expresses or denotes the amount of stretching necessary may conveniently be called the *tensor* of AB with respect to CD. (Hamilton.)

As far as two segments are concerned, the tensor, as a numerical quantity, is identical with the ratio of the segments, but it introduces a different idea. Hence in the case of commensurable segments the tensor is arithmetically expressible, but in the case of incommensurable ones the tensor may be symbolically denoted, but cannot be numerically expressed except approximately.

Thus if AB is the diagonal of a square of which CD is the side, $AB = CD\sqrt{2}$ (180°); and the tensor of AB on CD, *i.e.*, the measure of AB with CD as unit-length, is that numerical quantity which is symbolized by $\sqrt{2}$, and which can be expressed to any required degree of approximation by that arithmetical process known as "extracting the square root of 2."

190°. That the tensor symbolized by $\sqrt{2}$ cannot be expressed arithmetically is readily shown as follows:—

If $\sqrt{2}$ can be expressed numerically it can be expressed as a fraction, $\frac{m}{n}$, which is in its lowest terms, and where accordingly m and n are not both even.

If possible then let $\sqrt{2} = \frac{m}{n}$.

Then $2n^2 = m^2$. Therefore m^2 and m are both even and n is odd.

But if m is even, $\frac{m^2}{2}$ is even, and n^2 and n are both even.

But n cannot be both odd and even.

Therefore $\sqrt{2}$ cannot be arithmetically expressed.

Illustration of an incommensurable tensor.

Let BD be equal to AB, and let AC be equal to the diagonal of a square of which AB is the side.

```
                        E'F'
A───────────────B────E─C─F──────────D
```

Then *some* tensor will bring AB to AC.

Let BD be divided into 10 equal parts whereof E and F are those numbered 4 and 5.

Then the tensor 1.4 stretches AB to AE, and tensor 1.5 stretches AB to AF. But the first of these is too small and the second too great, and C lies between E and F.

Now, let EF be divided into 10 equal parts whereof E', F' are those numbered 1 and 2.

Then, tensor 1.41 brings AB to AE', and tensor 1.42 brings AB to AF'; the first being too small and the second too great.

Similarly by dividing E'F' into 10 equal parts we obtain two points *e*, *f*, numbered 4 and 5, which lie upon opposite sides of C and adjacent to it.

Thus, however far this process be carried, C will always lie *between* two adjacent ones of the points *last* obtained.

But as every new division gives interspaces one-tenth of the length of the former ones, we may obtain a point of division lying as near C as we please.

Now if AB be increased in length from AB to AD it must at some period of its increase be equal to AC.

Therefore the tensor which brings AB to AC is a real tensor which is inexpressible, except approximately, by the symbols of Arithmetic.

The preceding illustrates the difference between magnitude and number. The segment AB in changing to AD passes through *every* intermediate length. But the commensurable or numerically expressible quantities lying between 1 and 2 must proceed by some unit however small, and are therefore not continuous.

Hence *a magnitude is a variable which, in passing from one value to another, passes through every intermediate value.*

191°. The tensor of the segment AB with respect to AC, or the tensor of AB *on* AC is the numerical factor which brings AC to AB.

But according to the operative principles of Algebra,
$$\frac{AB}{AC} \cdot AC = AB,$$
$\therefore \frac{AB}{AC}$ is the tensor which brings AC to AB.

Hence *the algebraic form of a fraction, when the parts denote segments, is interpreted geometrically by the tensor which brings the denominator to the numerator; or as the ratio of the numerator to the denominator.*

SECTION I.

PROPORTION AMONGST LINE-SEGMENTS.

192°. *Def.—Four line-segments taken in order form a proportion, or are in proportion, when the tensor of the first on the second is the same as the tensor of the third on the fourth.*

This definition gives the relation
$$\frac{a}{b} = \frac{c}{d} \dots\dots\dots\dots\dots\dots(A)$$
where a, b, c, and d denote the segments taken in order.

The fractions expressing the proportion are subject to all the transformations of algebraic fractions (158°), and the result is geometrically true whenever it admits of a geometric interpretation.

The statement of the proportion is also written
$$a : b = c : d, \dots\dots\dots\dots\dots\dots(B)$$

PROPORTION AMONGST LINE-SEGMENTS. 151

where the sign : indicates the division of the quantity denoted by the preceding symbol by the quantity denoted by the following symbol.

In either form the proportion is read
"a is to b as c is to d."

193°. In the form (B) a and d are called the *extremes*, and b and c the *means*; and in both forms a and c are called *antecedents* and b and d *consequents*.

In the form (A) a and d, as also b and c, stand opposite each other when written in a cross, as
$$\frac{a \mid c}{b \mid d},$$
and we shall accordingly call them the opposites of the proportion.

194°. 1. From form (A) we obtain by cross-multiplication
$$ad = bc,$$
which states geometrically that

When four segments are in proportion the rectangle upon one pair of opposites is equal to that upon the other pair of opposites.

Conversely, let ab and $a'b'$ be equal rectangles having for adjacent sides a, b, and a', b' respectively. Then
$$ab = a'b',$$
and this equality can be expressed under any one of the following forms, or may be derived from any one of them, viz.:
$$\frac{a}{a'} = \frac{b'}{b}, \quad \frac{a}{b'} = \frac{a'}{b}, \quad \frac{b}{a'} = \frac{b'}{a}, \quad \frac{b}{b'} = \frac{a'}{a},$$
in all of which the opposites remain the same. Therefore

2. Two equal rectangles have their sides in proportion, a pair of opposites of the proportion coming from the same rectangle.

3. A given proportion amongst four segments may be written in any order of sequence, provided the opposites remain the same.

195°. The following transformations are important.

Let $\dfrac{a}{b} = \dfrac{c}{d}$, then

1. $\dfrac{a \pm b}{b} = \dfrac{c \pm d}{d}$, ($a > b$ for $-$ sign)

2. $\dfrac{a}{b} = \dfrac{c}{d} = \dfrac{a+c}{b+d} = \dfrac{a-c}{b-d}$. ($a > c$ for $-$ sign)

Let $\dfrac{a}{b} = \dfrac{c}{d} = \dfrac{e}{f} =$ etc., then

3. $\dfrac{a}{b} = \dfrac{c}{d} = \dfrac{e}{f} = \dfrac{a+c+e+\text{etc.}}{b+d+f+\text{etc.}}$.

To prove 1. $\because \dfrac{a}{b} = \dfrac{c}{d}$, $\therefore \dfrac{a}{b} \pm 1 = \dfrac{c}{d} \pm 1$, and $\dfrac{a \pm b}{b} = \dfrac{c \pm d}{d}$.

To prove 2. $\because \dfrac{a}{b} = \dfrac{c}{d}$, $\therefore \dfrac{a}{c} = \dfrac{b}{d}$, (194°, 3)

$\therefore \dfrac{a \pm c}{c} = \dfrac{b \pm d}{d}$,

or $\dfrac{a}{b} = \dfrac{c}{d} = \dfrac{a \pm c}{b \pm d}$.

To prove 3. $\because \dfrac{a}{b} = \dfrac{e}{f} = \dfrac{a+c}{b+d} = \dfrac{P}{Q}$, say,

$\therefore \dfrac{a}{b} = \dfrac{e}{f} = \dfrac{c+P}{f+Q} = \dfrac{a+c+e}{b+d+f}$, etc.

SIMILAR TRIANGLES.

196°. *Def.*—1. Two triangles are *similar* when the angles of the one are respectively equal to the angles of the other.

(77°, 4)

2. The sides opposite equal angles in the two triangles are corresponding or *homologous* sides.

The symbol $\backsimeq$ will be employed to denote similarity, and will be read "is similar to."

PROPORTION AMONGST LINE-SEGMENTS. 153

In the triangles ABC and A'B'C', if ∠A=∠A' and ∠B=∠B', then also ∠C=∠C' and the triangles are similar.

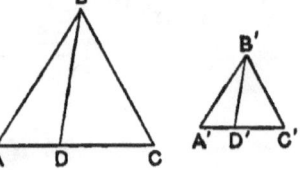

The sides AB and A'B' are homologous, so also are the other pairs of sides opposite equal angles.

Let BD through B and B'D' through B' make the
$$\angle BDA = \angle B'D'A'.$$

Then △ABD ∽ △A'B'D' since their angles are respectively equal. In like manner △DBC ∽ △D'B'C', and BD and B'D' divide the triangles similarly.

3. Lines which divide similar triangles similarly are homologous lines of the triangles, and the intersections of homologous *lines* are homologous *points*.

Cor. Evidently the perpendiculars upon homologous sides of similar triangles are homologous lines. So also are the medians to homologous sides; so also the bisectors of equal angles in similar triangles; etc.

197°. *Theorem.*—The homologous sides of similar triangles are proportional.

△ABC ∽ △A'B'C'

having ∠A=∠A'

and ∠B=∠B'.

Then $\dfrac{AB}{A'B'} = \dfrac{BC}{B'C'} = \dfrac{CA}{C'A'}.$

Proof.—Place A' on A, and let C' fall at D. Then, since ∠A'=∠A, A'B' will lie along AC and B' will fall at some point E. Now, △A'B'C' ≡ △AED, and therefore ∠AED=∠B, and B, D, E, C are concyclic. (107°)

Hence AD . AB = AE . AC, (176°, 2)

or A'C' . AB = A'B' . AC.

∴ $\dfrac{AB}{A'B'} = \dfrac{AC}{A'C'}.$ (194°, 2)

Similarly, by placing B' at B, we prove that

$$\frac{AB}{A'B'} = \frac{BC}{B'C'}.$$

∴ $$\frac{AB}{A'B'} = \frac{BC}{B'C'} = \frac{CA}{C'A'}.$$ q.e.d.

Cor. 1. Denoting the sides of ABC by a, b, c, and those of A'B'C' by a', b', c', $\quad \dfrac{a}{a'} = \dfrac{b}{b'} = \dfrac{c}{c'}.$

Cor. 2. $\quad \dfrac{a}{a'} = \dfrac{b}{b'} = \dfrac{c}{c'} = \dfrac{a+b+c}{a'+b'+c'},\quad$ (195°, 3)

i.e., the perimeters of similar triangles are proportional to any pair of homologous sides.

198°. Theorem.—Two triangles which have their sides proportional are similar, and have their equal angles opposite homologous sides. (Converse of 197°.)

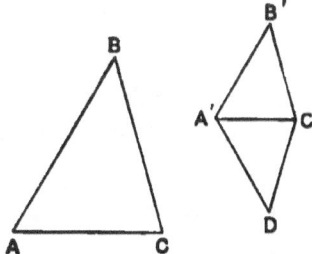

$$\frac{AB}{A'B'} = \frac{BC}{B'C'} = \frac{CA}{C'A'}.$$

Then ∠A=∠A', ∠B=∠B', and ∠C=∠C'.

Proof.—On A'C' let the △A'DC' be constructed so as to have the ∠DA'C'=∠A
and ∠DC'A'=∠C.
Then △A'DC'∽△ABC, (196°, Def. 1)
and $\dfrac{AB}{A'D} = \dfrac{AC}{A'C'} = \dfrac{BC}{DC'}$; (197°)
but $\dfrac{AB}{A'B'} = \dfrac{AC}{A'C'} = \dfrac{BC}{B'C'}$; (hyp.)
∴ A'D=A'B'
and DC'=B'C',
and △A'DC'≡△A'B'C'. (58°)
∴ ∠A'=∠A, ∠B'=∠B,
and ∠C'=∠C. q.e.d.

PROPORTION AMONGST LINE-SEGMENTS.

199°. Theorem.—If two triangles have two sides in each proportional and the included angles equal, the triangles are similar.

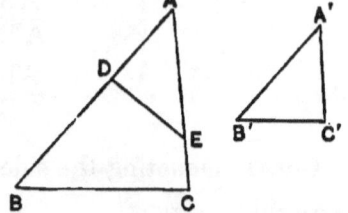

$\dfrac{AB}{A'B'} = \dfrac{AC}{A'C'}$, and $\angle A = \angle A'$,

then $\triangle ABC \backsim \triangle A'B'C'$.

Proof.—Place A' on A, and let $A'C'$ lie along AB, and $A'B'$ lie along AC, so that C' falls at D and B' at E.

The triangles AED and $A'B'C'$ are congruent and therefore similar, and $\dfrac{AB}{AE} = \dfrac{AC}{AD}$

Hence $AB \cdot AD = AE \cdot AC$; (194°)
and $\therefore$ B, D, E, C are concyclic. (177°)
$\therefore \quad \angle AED = \angle B$, and $\angle ADE = \angle C$, (106°, Cor. 3)
and $\quad \triangle ABC \backsim \triangle AED \backsim \triangle A'B'C'$. q.e.d.

200°. Theorem.—If two triangles have two sides in each proportional, and an angle opposite a homologous side in each equal:

1. If the angle is opposite the longer of the two sides the triangles are similar.

2. If the angle is opposite the shorter of the two sides the triangles may or may not be similar.

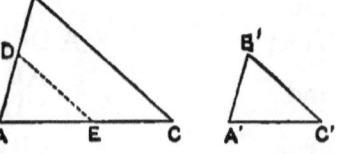

$\dfrac{AB}{A'B'} = \dfrac{BC}{B'C'}$, and $\angle A = \angle A'$.

1. If $BC > AB$,
$\triangle ABC \backsim \triangle A'B'C'$.

Proof.—Place A' at A and let B' fall at D, and $A'C'$ along AC. Draw $DE \parallel$ to BC. Then
$\triangle ABC \backsim \triangle ADE$, and $\dfrac{AB}{AD} = \dfrac{BC}{DE}$.

But $\dfrac{AB}{AD} = \dfrac{BC}{B'C'}$; $\therefore DE = B'C'$.

And since B'C' > A'B', the △A'B'C' ≡ △ADE and they are therefore similar. (65°, 1)
But △ABC ≎ △ADE,
∴ △ABC ≎ △A'B'C'.

2. If BC < AB, B'C' < A'B', and the triangles may or may not be similar.

Proof.—Since AD = A'B', and DE = B'C', and B'C' < A'B', ∴ the triangles A'B'C' and ADE may or may not be congruent (65°, 2), and therefore may or may not be similar.
But △ABC ≎ △ADE,
∴ the triangles ABC and A'B'C' may or may not be similar.

Cor. Evidently, if in addition to the conditions of the theorem, the angles C and C' are both less, equal to, or greater than a right angle the triangles are similar.

Also, if the triangles are right-angled they are similar.

201°. The conditions of similarity of triangles may be classified as follows :—

1. Three angles respectively equal. (Def. of similarity.)
2. Three sides proportional.
3. Two sides proportional and the included angles equal.
4. Two sides proportional and the angles opposite the longer of the homologous sides in each equal.

If in 4 the equal angles are opposite the shorter sides in each the triangles are not necessarily similar unless some other condition is satisfied.

By comparing this article with 66° we notice that there is a manifest relation between the conditions of congruence and those of similarity.

Thus, if in 2, 3, and 4 of this article the words "proportional" and "homologous" be changed to "equal," the statements become equivalent to 1, 2, and 5 of Art. 66°. The difference between congruence and similarity is the non-necessity of equality of areas in the latter case.

When two triangles, or other figures, are similar, they are copies of one another, and the smaller may be brought, by a uniform stretching of all its parts, into congruence with the larger. Thus the primary idea of similarity is that every line-segment of the smaller of two similar figures is stretched to the same relative extent to form the corresponding segments of the larger figure. This means that the tensors of every pair of corresponding line-segments, one from each figure, are equal, and hence that any two or more line-segments from one figure are proportional to the corresponding segments from the second figure.

Def.—Two line-segments are *divided similarly* when, being divided into the same number of parts, any two parts from one of the segments and the corresponding parts from the other taken in the same order are in proportion.

202°. *Theorem.*—A line parallel to the base of a triangle divides the sides similarly; and

Conversely, a line which divides two sides of a triangle similarly is parallel to the third side.

DE is ∥ to AC. Then BA and BC are divided similarly in D and E.

Proof.—The triangles ABC and DBE are evidently similar,

∴ $\dfrac{AB}{DB} = \dfrac{CB}{EB}$, and ∴ $\dfrac{AD}{DB} = \dfrac{CE}{EB}$, (195°, 1)

and AB and CB are divided similarly in D and E. *q.e.d.*

Conversely, if DE so divides BA and BC that
AD : DB = CE : EB, DE is ∥ to AC.

Proof.—Since $\dfrac{AD}{DB} = \dfrac{CE}{EB}$, ∴ $\dfrac{AB}{DB} = \dfrac{CB}{EB}$; and the triangles ABC and DBE having the angle B common, and the sides

158   SYNTHETIC GEOMETRY.

about that angle proportional, are similar.   (199°)
∴ $\angle BDE = \angle A$, and DE is ∥ to AC.   *q.e.d.*

Cor. 1. Since the triangles ABC and DBE are similar
$$BA : BD = AC : DE.$$

203°. *Theorem.*—Two transversals to a system of parallels are divided similarly by the parallels.
AA′ is ∥ to BB′ is ∥ to CC′, etc.
Then AD and A′D′ are divided similarly.

Proof.—Consider three of the ∥s, AA′, BB′, and CC′, and draw A′Q ∥ to AD.
Then AP and BQ are ▱s, and
∴   AB = A′P and BC = PQ. (81°, 1)
But A′QC′ is a triangle and PB′ is ∥ to QC′.

∴ $\dfrac{A'P}{PQ} = \dfrac{A'B'}{B'C'}$ (202°), or $\dfrac{AB}{A'B'} = \dfrac{BC}{B'C'}$ (194°, 3).

Similarly, if DD′ be a fourth parallel, $\dfrac{BC}{B'C'} = \dfrac{CD}{C'D'}$

∴ $\dfrac{AB}{A'B'} = \dfrac{BC}{B'C'} = \dfrac{CD}{C'D'} =$ etc.

Def.—A set of three or more lines meeting in a point is a *pencil* and the lines are *rays*.
The point is the *vertex* or *centre* of the pencil.

Cor. 1. Let the transversals meet in O, and let L denote any other transversal through O.
Then AD, A′D′, and L are all divided similarly by the parallels. But the parallels are transversals to the pencil.
∴ parallel transversals divide the rays of a pencil similarly.

Cor. 2. Applying Cor. 1 of 202°,
$$\dfrac{OA}{AA'} = \dfrac{OB}{BB'} = \dfrac{OC}{CC'} = \dfrac{OD}{DD'} = \text{etc.}$$

PROPORTION AMONGST LINE-SEGMENTS. 159

204°. *Theorem.*—The rectangle on any two sides of a triangle is equal to twice the rectangle on the circumradius (97°, Def.) and the altitude to the third side.

BD is $\perp$ to AC and BE is a diameter.
Then $\quad$ BA.BC = BE.BD.

Proof.— $\quad \angle A = \angle E,\quad$ (106°, Cor. 1)
and $\angle ADB = \angle ECB = \rceil,\quad$ (106°, Cor. 4)

$\therefore \quad \triangle ABD \backsim \triangle EBC,$ and $\dfrac{BA}{BE} = \dfrac{BD}{BC},$

$\therefore \quad$ BA.BC = BE.BD. $\qquad$ *q.e.d.*

Cor. Denoting BD by p and the circumradius by R,
$$ac = 2pR,$$
and multiplying by b, and remembering that $pb = 2\triangle$ (153°, 2),
we obtain $\quad R = \dfrac{abc}{4\triangle},$

which (with 175½°, Ex. 1) gives the means of calculating the circumradius of a triangle when its three sides are given.

205°. *Theorem.*—In a concyclic quadrangle the rectangle on the diagonals is equal to the sum of the rectangles on the sides taken in opposite pairs.

AC.BD = AB.CD + BC.AD.

Proof.—Draw AE making $\angle AED = \angle ABC$. Then, since $\angle BCA = \angle BDA$, the triangles EDA and BCA are similar.

$\therefore \quad$ BC.AD = AC.DE.

Again, since $\angle AEB$ is supp. to $\angle AED$, and $\angle CDA$ is supp. to $\angle ABC$, therefore triangles BEA and CDA are similar, and AB.CD = AC.EB.

Adding these results, AB.CD + BC.AD = AC.BD.

This theorem is known as *Ptolemy's* Theorem.

206°. *Def.*—Two rectilinear figures are similar when they can be divided into the same number of triangles similar in pairs and similarly placed.

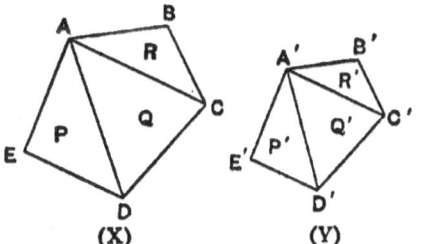

Thus the pentagons X and Y can be divided into the same number of triangles.

If then $\triangle P \backsimeq \triangle P'$, $\triangle Q \backsimeq \triangle Q'$, $\triangle R \backsimeq \triangle R'$, and the triangles are similarly placed, the pentagons are similar.

The triangles are similarly placed if ∠EAD corresponds to E'A'D', ∠AED to A'E'D', ∠DAC to D'A'C', etc.

This requires that the angles A, B, C, etc., of one figure shall be respectively equal to the angles A', B', C', etc., of the other figure.

Hence when two rectilinear figures are similar, their angles taken in the same order are respectively equal, and the sides about equal angles taken in the same order are proportional.

Line-segments, such as AD and A'D', which hold similar relations to the two figures are similar or homologous lines of the figures.

207°. *Theorem*.—Two similar rectilinear figures have any two line-segments from the one proportional to the homologous segments from the other.

Proof.—By definition $\triangle P \backsimeq \triangle P'$, and they are similarly placed, ∴ AE : A'E' = AC : A'C'.
For like reasons, AD : A'D' = AC : A'C' = AB : A'B'.

∴ $\dfrac{AE}{A'E'} = \dfrac{AD}{A'D'} = \dfrac{AC}{A'C'} = \dfrac{AB}{A'B'} =$ etc.,

and the same can be shown for any other sets of homologous line-segments.

Cor. 1. All regular polygons of the same species are similar figures.

PROPORTION AMONGST LINE-SEGMENTS. 161

Now, let $a, b, c, \ldots, a', b', c', \ldots$, be homologous sides of two similar regular polygons, and let r and r' be their circumradii.

Then r and r' are homologous,

$$\therefore \quad \frac{r}{r'} = \frac{a}{a'} = \frac{b}{b'} = \frac{c}{c'} = \ldots = \frac{a+b+c+\ldots}{a'+b'+c'+\ldots} \quad (195°, 3)$$

$$= \frac{\text{perimeter of P}}{\text{perimeter of P'}}.$$

But at the limit (148°) the polygon becomes its circumcircle.

∴ the circumferences of any two circles are proportional to their radii.

Cor. 2. If c, c' denote the circumferences of two circles and r and r' their radii, $\quad \dfrac{c}{r} = \dfrac{c'}{r'} = $ constant.

Denote this constant by 2π, then

$$c = 2\pi r.$$

It is shown by processes beyond the scope of this work that π stands for an incommensurable numerical quantity, the approximate value of which is 3.1415926...

Cor. 3. Since equal arcs subtend equal angles at the centre (102°, Cor. 2), if s denotes the length of any arc of a circle whose radius is r, the tensor $\dfrac{s}{r}$ varies directly as s varies, and also varies directly as the angle at the centre varies.

Hence $\dfrac{s}{r}$ is taken as the *measure* of the angle, subtended by the arc, at the centre. Denote this angle by θ. Then

$$\theta = \frac{s}{r},$$

and when $s = r$, θ becomes the unit angle.

∴ the unit angle is the angle subtended at the centre by an arc equal in length to the radius.

This unit is called a *radian*, and the measure of an angle in radians is called its radian measure. Radian measure will be indicated by the mark ^.

Cor. 4. When $s = \dfrac{c}{2} = $ a semicircle, $\theta = \pi$.

L

But a semicircle subtends a straight angle at the centre.

∴ π is the radian measure of a straight angle and $\dfrac{\pi}{2}$ of a ⌐.

Now a straight angle contains 180°, (41°)

∴ $\pi^\wedge = 180°$.

Hence $1^\wedge = 57°.29578\ldots$,

and $1° = 0^\wedge.017453\ldots$;

and these multipliers serve to change the expression of a given angle from radians to degrees or from degrees to radians.

Cor. 5. Since the area of a circle is equal to one-half that of the rectangle on its radius and a segment equal in length to its circumference, (149°)

∴ $\odot = \tfrac{1}{2} cr = \tfrac{1}{2} \cdot 2\pi r \cdot r$ (Cor. 2)

$\qquad = \pi r^2$.

∴ the area of a $\odot$ is π times that of the square on its radius.

208°. Theorem.—The bisectors of the vertical angle of a triangle each divides the base into parts which are proportional to the conterminous sides.

BD and BD′ are bisectors of ∠B. Then

$$\dfrac{AD}{DC} = \dfrac{AD'}{CD'} = \dfrac{AB}{BC}.$$

Proof.—Through C draw EE′ ∥ to AB. Then
EBE′ = ⌐ (45°), and ∠E = ∠ABD = ∠DBC.

∴ BC = EC = CE′. (88°, 3)

But ABD and ABD′ are triangles having EE′ ∥ to the common base AB.

∴ $\dfrac{AB}{EC} = \dfrac{AD}{DC}$ and $\dfrac{AB}{CE'} = \dfrac{AD'}{CD'}$, (203°, Cor.)

or $\dfrac{AD}{DC} = \dfrac{AD'}{CD'} = \dfrac{AB}{BC}$. *q.e.d.*

Cor. D and D′ divide the base internally and externally in the same manner. Such division of a segment is called *harmonic* division.

∴ the bisectors of any angle of a triangle divide the opposite side harmonically.

209°. *Theorem.* — A line through the vertex of a triangle dividing the base into parts which are proportional to the conterminous sides is a bisector of the vertical angle. (Converse of 208°.)

Let the line through B cut AC internally in F. Then, AD being the internal bisector $\frac{AB}{BC} = \frac{AD}{DC}$ (208°), and $\frac{AB}{BC} = \frac{AF}{FC}$ by hypothesis, ∴ $\frac{AF}{FC} = \frac{AD}{DC}$.

But AD is < AF while DC is > FC.

∴ the relation is impossible unless F and D coincide, *i.e.*, the line is the bisector AD.

Similarly it may be proved that if the line divides the base externally it is the bisector AD'.

210°. *Theorem.* — The tangent at any point on a circle and the perpendicular from that point upon the diameter divide the diameter harmonically.

AB is divided harmonically in M and T.

Proof. — ∠CPT = ∠PMT = ⌐, (110°)

∴ △CPM ≃ △CTP,

and $\frac{CM}{CP} = \frac{CP}{CT}$, or $\frac{CM}{CB} = \frac{CB}{CT}$;

∴ $\frac{CB + CM}{CB - CM} = \frac{CT + CB}{CT - CB}$, or $\frac{AM}{MB} = \frac{AT}{BT}$, (195°, 2)

∴ AB is divided harmonically in M and T.   *q.e.d.*

211°. The following examples give important results.

Ex. 1. L, M, and N are tangents which touch the circle at A, B, and P.

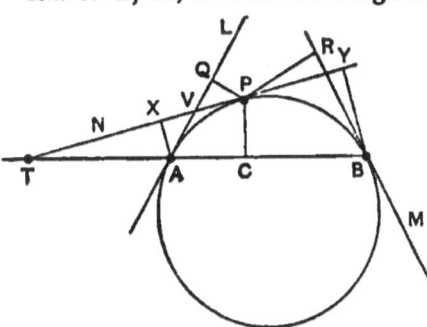

AX and BY are ⊥s on N, PC is ⊥ on AB, and PQ and PR are ⊥s upon L and M.

Let N meet the chord of contact of L and M in T. Then the triangles TAX, TPC, TBY are all similar, ∴ $\frac{TA}{AX} = \frac{TP}{PC} = \frac{TB}{BY}$; ∴ $\frac{TA \cdot TB}{AX \cdot BY} = \frac{TP^2}{PC^2}$.

But $\quad TA \cdot TB = TP^2,\quad$ (176°, Cor. 2)
∴ $\quad AX \cdot BY = PC^2$. ....(A)

Again, let L and N intersect in V. Then
$$VP = VA, \quad (114°, \text{Cor. 1})$$
$$\angle VQP = \angle VXA = \sqcap,$$
and $\quad \angle QVP = \angle XVA.$
∴ $\quad \triangle VXA \equiv \triangle VQP,$
and $\quad AX = PQ.$
Similarly $\quad BY = PR,\; ∴\; PQ \cdot PR = PC^2$. ....(B)

Ex. 2. AD is a centre-line and DQ a perpendicular to it, and AQ is any line from A to the line DQ.

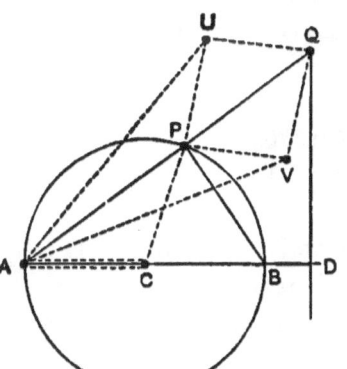

Let AQ cut the circle in P. Then $\triangle ADQ \approx \triangle APB,$
∴ $\quad \frac{AD}{AP} = \frac{AQ}{AB},$
or $\quad AD \cdot AB = AP \cdot AQ.$
But the circle and the point D being given, $AB \cdot AD$ is a given constant.
∴ $\quad AP \cdot AQ = $ a constant.

Conversely, if Q moves so that the ▭$AP \cdot AQ$ remains constant, the locus of Q is a line ⊥ to the centre-line through A.

Now, let the dotted lines represent rigid rods of wood or metal jointed together so as to admit of free rotation about the points A, C, P, U, V, and Q, and such that UPVQ is a rhombus (82°, Def. 1), and AU=AV, and AC=CP, AC being fixed.

PQ is the right bisector of UV, and A is equidistant from U and V. Therefore A, P, Q are always in line.

Also, PUQ is an isosceles triangle and UA is a line to the base, therefore $UA^2 - UP^2 = AP . AQ$ (174°). But, UA and UP being constants, AP.AQ is constant.

And AC being fixed, and CP being equal to AC, P moves on the circle through A having C as centre.

∴ Q describes a line ⊥ to AC.

This combination is known as *Peaucellier's cell*, and is interesting as being the first successful attempt to describe a line by circular motions only.

Ex. 3. To construct an isosceles triangle of which each basal angle shall be double the vertical angle.

Let ABC be the triangle required, and let AD bisect the ∠A.

Then ∠B=∠BAD=∠DAC, and ∠C is common to the triangles ABC and DAC. Therefore these triangles are similar, and the △CAD is isosceles and AD=AC.

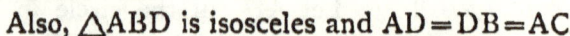

Also, △ABD is isosceles and AD=DB=AC.
∴   BA : AC = AC : DC,
or   BC : BD = BD : DC.
∴ $BC . DC = BD^2$.

And BC is divided into extreme and mean ratio at D (183°, Ex. 1). Thence the construction is readily obtained.

Cor. 1. The isosceles triangle ADB has each of its basal angles equal to one-third its vertical angle.

Cor. 2. ∠ABC=36°, ∠BAC=72°, ∠BDA=108°. Hence
(1) Ten triangles congruent with ABC, placed side by side with their vertices at B, form a regular decagon. (132°)

(2) The bisectors AD and CE of the basal angles of the △ABC meet its circumcircle in two points which, with the three vertices of the triangle, form the vertices of a regular pentagon.

(3) The ∠BDA = the internal angle of a regular pentagon.

211½°. The following Mathematical Instruments are important:—

1. *Proportional Compasses.*

This is an instrument primarily for the purpose of increasing or diminishing given line-segments in a given ratio; *i.e.*, of multiplying given line-segments by a given tensor.

If AO = BO and QO = PO, the triangles AOB, POQ are isosceles and similar, and

AB : PQ = OA : OP.

Hence, if the lines are one or both capable of rotation about O, the distance AB may be made to vary at pleasure, and PQ will remain in a constant ratio to AB.

The instrument usually consists of two brass bars with slots, exactly alike, and having the point of motion O so arranged as to be capable of being set at any part of the slot. The points A, B, P, and Q are of steel.

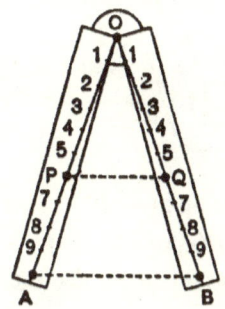

2. *The Sector.*

This is another instrument which primarily serves the purpose of increasing or diminishing given line-segments in given ratios.

This instrument consists of two rules equal in length and jointed at O so as to be opened and shut like a pair of compasses. Upon each rule various lines are drawn corresponding in pairs, one on each rule.

Consider the pair OA and OB, called the "line of lines." Each of the lines of this pair is divided into 10 equal parts

which are again subdivided. Let the divisions be numbered from 0 to 10 along OA and OB, and suppose that the points numbered 6 are the points P and Q. Then OAB and OPQ are similar triangles, and therefore $PQ:AB = OP:OA$. But $OP = \frac{6}{10}AO$. $\therefore$ $PQ = \frac{6}{10}AB$.
And as by opening the instrument AB may be made equal to any segment not beyond the compass of the instrument, we can find PQ equal to $\frac{6}{10}$ of any such given segment.

The least consideration will show that the distance 5-5 is $\frac{1}{2}AB$, 3-3 is $\frac{3}{10}AB$, etc. Also that 3-3 is $\frac{3}{7}$ of 7-7, 5-5 is $\frac{5}{7}$ of 7-7, etc. Hence the instrument serves to divide any given segment into any number of equal parts, provided the number is such as belongs to the instrument.

The various other lines of the sector serve other but very similar purposes.

3. *The Pantagraph or Eidograph.*

Like the two preceding instruments the pantagraph primarily increases or diminishes segments in a given ratio, but unlike the others it is so arranged as to be continuous in its operations, requiring only one setting and no auxiliary instruments.

It is made of a variety of forms, but the one represented in the figure is one of the most convenient.

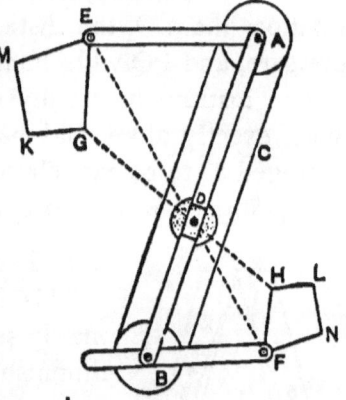

AE, AB, and BF are three bars jointed at A and B. The bars AE and BF are attached to the wheels A and B respectively, which are exactly of the same diameter, and around which goes a very thin and flexible steel band C.

The result is that if AE and BF are so adjusted as to be parallel, they remain parallel however they be situated with respect to AB. E, F are two points adjustable on the bars

AE and BF, and D is a point in line with EF, around which the whole instrument can be rotated.

Now let EGKM be any figure traced by the point E; then F will trace a similar figure FHLN.

Evidently the triangles DAE and DBF remain always similar however the instrument is transformed. Therefore DF is in a constant ratio to DE, viz., the ratio DB : DA.

Now, when E comes to G, F comes to some point H in line with GD, and such that DH : DG = DB : DA.

∴ the triangles EDG and FDH are similar, and FH is ∥ to EG, and has to it the constant ratio DB : DA. Similarly HL is ∥ to GK and has to it the same constant ratio, etc.

∴ the figures are similar, and the ratio of homologous lines in GM and HN is AD : DB.

The three points E, D, and F being all adjustable the ratio can be changed at pleasure.

Altogether the Pantagraph is a highly important instrument, and when so adjusted that E, D, and F are not in line its results offer some interesting geometrical features.

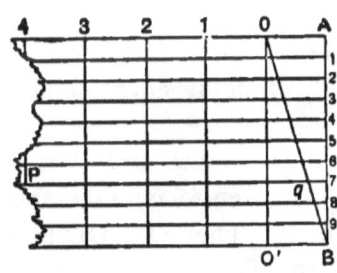

4. *The Diagonal Scale.*

This is a divided scale in which, by means of similar triangles, the difficulty of reading off minute divisions is very much diminished.

Its simpler form is illustrated in the figure.

A scale divided to fortieths of an inch is, on account of the closeness of the divisions, very difficult to read.

In the scale represented OA is ¼ inch. The distance AB is divided into 10 equal parts by the horizontal parallel lines numbered 1, 2, 3, etc. Then OBO' is a triangle whereof the horizontal lines are all parallel to the base. Hence it is readily seen from the proportionality of the homologous sides of the similar triangles formed that the intercept on the horizontal line 1, between OO′ and OB, is $\frac{1}{10}$O′B, that is $\frac{1}{40}$ inch.

Similarly the intercept on the horizontal line 2 is $\frac{2}{40}$ inch, on 3, $\frac{3}{40}$ inch, etc.

Hence from p to q is one inch and seven-fortieths.

In a similar manner diagonal scales can be made to divide any assumed unit-length into any required number of minute parts.

The chief advantages of such scales are that the minute divisions are kept quite distinct and apparent, and that errors are consequently avoided.

Exercises.

1. ABCD is a square and P is taken in BC so that PC is one-third of BC. AC cuts the diagonal BD in O, and AP cuts it in E. Then OE is one-tenth of DB.
2. If, in 1, OE is one-eighth of DB, how does P divide BC?
3. If BP is one nth of BC, what part of DB is OE?
4. Given three line-segments to find a fourth, so that the four may be in proportion.
5. The rectangle on the distances of a point and its chord of contact from the centre of a circle is equal to the square on the radius of the circle.
6. OD and DQ are fixed lines at right angles and O is a fixed point. A fixed circle with centre on OD and passing through O cuts OQ in P. Then OP . OQ is a constant however OQ be drawn.
7. To divide a given segment similarly to a given divided segment.
8. To divide a given segment into a given number of equal parts.
9. Two secants through A cut a circle in B, D, and C, E respectively. Then the triangles ABE and ACD are similar. So also are the triangles ABC and AED.
10. Two chords are drawn in a circle. To find a point on

the circle from which perpendiculars to the chords are proportional to the lengths of the chords.

11. ABC is a triangle and DE is parallel to AC, D being on AB and E on CB. DC and AE intersect in O. Then BO is a median.
12. If BO, in 11, cuts DE in P and AC in Q, BO is divided harmonically by P and Q.
13. A and B are centres of fixed circles and AX and BY are parallel radii. Show that XY intersects AB in a fixed point.
14. In the triangle ABC, BD bisects the $\angle$B and cuts AC in D. Then $BD^2 = AB \cdot BC - AD \cdot DC$. (Employ the circumcircle.)
15. ABC is right-angled at B and BD is the altitude on AC.
    (1) The $\triangle$s ADB and BDC are each similar to ABC.
    (2) Show by proportion that $AB^2 = AD \cdot AC$,
    and $BD^2 = AD \cdot DC$.
16. If R and r denote the radii of the circumcircle and incircle of a triangle, $2Rr(a+b+c) = abc$.
17. In an equilateral triangle the square on the side is equal to six times the rectangle on the radii of the circumcircle and incircle.
18. OA, OB, OC are three lines. Draw a line cutting them so that the segment intercepted between OA and OC may be bisected by OB.
19. What is the measure of an angle in radians when its measure in degrees is 68° 17′?
20. How many radians are in the angle of an equilateral $\triangle$?
21. The earth's diameter being 7,960 miles, what is the distance in miles between two places having the same longitude but differing 16° in latitude?
22. Construct a regular pentagon, a regular decagon, a regular polygon—of 15 sides, of 30 sides, of 60 sides.
23. ABCDE is a regular pentagon.
    (1) Every diagonal is divided into extreme and mean ratio by another diagonal.

(2) The diagonals enclose a second regular pentagon.
24. Compare the side and the areas of the two pentagons of 23 (2).
25. If one side of a right-angled triangle is a mean proportional between the other side and the hypothenuse, the altitude from the right angle divides the hypothenuse into extreme and mean ratio.
26. A variable line from a fixed point A meets a fixed circle in P, and X is taken on AP so that $AP \cdot AX = $ a constant. The locus of X is a circle.
27. If two circles touch externally their common tangent is a mean proportional between their diameters.
28. Four points on a circle are connected by three pairs of lines. If a, a_1 denote the perpendiculars from any fifth point on the circle to one pair of lines, β, β_1 to another pair, and γ, γ_1 to the third pair, then $aa_1 = \beta\beta_1 = \gamma\gamma_1$. (Employ 204°.)
29. A line is drawn parallel to the base of a trapezoid and bisecting the non-parallel sides. Compare the areas of the two trapezoids formed.
30. Draw two lines parallel to the base of a triangle so as to trisect the area.
31. ABC is right-angled at B, and AP is the perpendicular from A to the tangent to the circumcircle at B. Then $AP \cdot AC = AB^2$.

SECTION II.

FUNCTIONS OF ANGLES.—AREAL RELATIONS.

212°. *Def.*—When an element of a figure undergoes change the figure is said to *vary* that element.

If a triangle changes into any similar triangle it varies its magnitude while its form remains constant; and if it changes into another form while retaining the same area, it varies its form while its area remains constant, etc.

Similar statements apply to other figures as well as triangles.

When a triangle varies its magnitude only, the tensors or ratios of the sides taken two and two remain constant. Hence the tensors or ratios of the sides of a triangle taken two and two determine the form of the triangle but not its magnitude; *i.e.*, they determine the angles but not the sides.

(77°, 3; 197°; 198°; 201°)

A triangle, which, while varying its size, retains its form, is sometimes said to remain similar to itself, because the triangles due to any two stages in its variation are similar to one another.

213°. In the right-angled triangle the ratios or tensors of the sides taken in pairs are important functions of the angles and receive distinctive names.

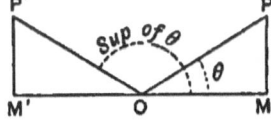

The $\triangle$OPM is right-angled at M, and the $\angle$POM is denoted by θ.

Then, $\dfrac{PM}{OP}$ is the *sine* of θ, and is contracted to $\sin \theta$ in writing.

$\dfrac{OM}{OP}$ is the sine of the $\angle$OPM, but as $\angle$OPM is the complement of θ, this tensor is called the *cosine* of θ, and is written $\cos \theta$.

$\dfrac{PM}{OM}$ is the *tangent* of θ, contracted to $\tan \theta$.

Cor. 1. $\because \quad \dfrac{PM}{OM} = \dfrac{PM}{OP} \cdot \dfrac{OP}{OM}, \quad \therefore \tan \theta = \dfrac{\sin \theta}{\cos \theta}$.

Cor. 2. $\because \quad PM^2 + OM^2 = OP^2, \quad \therefore \left(\dfrac{PM}{OP}\right)^2 + \left(\dfrac{OM}{OP}\right)^2 = 1$.
or $\sin^2 \theta + \cos^2 \theta = 1$.

214°. Let $OP' = OP$ be drawn so that $\angle P'OM' = POM = \theta$, and let $P'M'$ be $\perp$ on OM.

Then $\angle P'OM$ is the supplement of θ, and the triangles P'OM' and POM are congruent.

1. $$\sin P'OM = \frac{P'M'}{OP'} = \frac{PM}{OP} = \sin\theta.$$

i.e., an angle and its supplement have the same sine.

2. $\cos P'OM = \frac{OM'}{OP'}$. But in changing from OM to OM', on the same line, OM vanishes and then reappears upon the opposite side of O.

Therefore OM and OM' have opposite senses (156°), and if we consider OM positive, OM' is negative.

∴ $\qquad$ OM' = – OM, and hence

an angle and its supplement have cosines which are equal in numerical value but opposite in sign.

215°. *Theorem.* — The area of a parallelogram is the product of two adjacent sides multiplied by the sine of their included angle. (152°, 1)

AC is a ▱ and BP is ⊥ upon AD. Then BP is the altitude, and the area = AD . BP. (153°, 1)

But $\qquad$ BP = AB sin ∠BAP.

∴ $\qquad$ ▱ = AB . AD sin ∠BAP = $ab \sin\theta$.

Cor. 1. Since the area of a triangle is one-half that of the parallelogram on the same base and altitude,

∴ the area of a triangle is one-half the product of any two sides multiplied by the sine of the included angle. Or
$$2\triangle = ab \sin C = bc \sin A = ca \sin B.$$

216°. *Theorem.* — The area of any quadrangle is one-half the product of the diagonals multiplied by the sine of the angle between them.

ABCD is a quadrangle of which AC and BD are diagonals.

Let ∠AOB = θ = ∠COD.

Then ∠BOC = ∠AOD = supp. of θ.

$\triangle AOB = \tfrac{1}{2} OA . OB \sin\theta,$

$\triangle BOC = \tfrac{1}{2} OB . CO \sin\theta, \quad \triangle COD = \tfrac{1}{2} OC . DO \sin\theta,$

$\triangle \text{DOA} = \tfrac{1}{2} \text{DO} \cdot \text{OA} \sin \theta$, and adding,
   Qd. $= \tfrac{1}{2}$ AC . BD $\sin \theta$. (compare 162°)

217°. BD being the altitude to AC in the $\triangle$ABC, we have from 172°, 2,

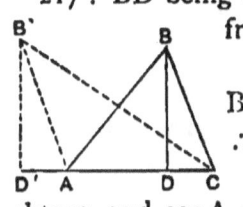

$$a^2 = b^2 + c^2 - 2b \cdot \text{AD}.$$
But AD $=$ AB cos A $= c$ cos A,
∴ $a^2 = b^2 + c^2 - 2bc \cos A.$

When B comes to B' the $\angle$A becomes obtuse, and cos A changes sign.   (214°, 2)

If we consider the cosine with respect to its magnitude only, we must write $+$ before the term $2bc \cos A$, when A becomes obtuse. But, if we leave the sign of the function to be accounted for by the character of the angle, the form given is universal.

Cor. 1. ABCD is a parallelogram. Consider the $\triangle$ABD, then $\text{BD}^2 = a^2 + b^2 - 2ab \cos \theta.$

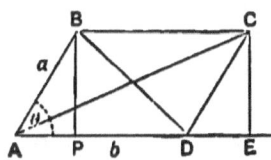

Next, consider the $\triangle$ABC. Since $\angle$ABC is the supplement of θ, and
  BC $=$ AD $= b$,
∴ $\text{AC}^2 = a^2 + b^2 + 2ab \cos \theta.$
and writing these as one expression,
 $a^2 + b^2 \pm 2ab \cos \theta,$
gives both the diagonals of any ▱, one of whose angles is θ.

Cor. 2. DE $= a \cos \theta$ (CE being $\perp$ to AD), CE $= a \sin \theta$.
∴ AE $= b + a \cos \theta$;
and $\tan \text{CAE} = \dfrac{\text{CE}}{\text{AE}} = \dfrac{a \sin \theta}{b + a \cos \theta},$
which gives the direction of the diagonal.

218°. *Def.*—The ratio of any area X to another area Y is the measure of X when Y is taken as the unit-area, and is accordingly expressed as $\dfrac{X}{Y}$. (Compare 188°.)

1. Let X and Y be two similar rectangles. Then X $= ab$

FUNCTIONS OF ANGLES.—AREAL RELATIONS.

and $Y = a'b'$, where a and b are adjacent sides of the $\square X$ and a' and b' those of the $\square Y$.

$$\therefore \quad \frac{X}{Y} = \frac{a}{a'} \cdot \frac{b}{b'}$$

But because the rectangles are similar, $\dfrac{a}{a'} = \dfrac{b}{b'}$

$$\therefore \quad \frac{X}{Y} = \frac{a^2}{a'^2},$$

i.e., the areas of similar rectangles are proportional to the areas of the squares upon homologous sides.

2. Let X and Y be two similar triangles. Then
$$X = \tfrac{1}{2}ab \sin C, \qquad Y = \tfrac{1}{2}a'b' \sin C,$$
$$\therefore \quad \frac{X}{Y} = \frac{ab}{a'b'} = \frac{a^2}{a'^2},$$
because the triangles are similar, \hfill (197°)
i.e., the areas of similar triangles are proportional to the areas of the squares upon homologous sides.

3. Let X denote the area of the pentagon ABCDE, and Y that of the similar pentagon A'B'C'D'E'. Then

$$\frac{P}{P'} = \frac{AD^2}{A'D'^2}, \quad \frac{R}{R'} = \frac{AC^2}{A'C'^2},$$
$$\frac{Q}{Q'} = \frac{DC^2}{D'C'^2}.$$

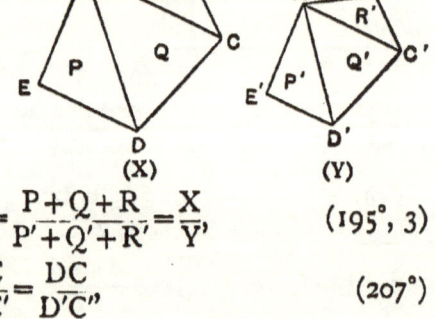

(X)　　(Y)

But
$$\frac{P}{P'} = \frac{R}{R'} = \frac{Q}{Q'} = \frac{P+Q+R}{P'+Q'+R'} = \frac{X}{Y}, \qquad (195°, 3)$$

and
$$\frac{AD}{A'D'} = \frac{AC}{A'C'} = \frac{DC}{D'C'} \qquad (207°)$$

$$\therefore \quad \frac{X}{Y} = \frac{DC^2}{D'C'^2}.$$

And the same relation may be proved for any two similar rectilinear figures whatever.

∴ the areas of any two similar rectilinear figures are proportional to the areas of squares upon any two homologous lines.

4. Since two circles are always similar, and are the limits of two similar regular polygons,

∴ the areas of any two circles are proportional to the areas of squares on any homologous chords of the circles, or on line-segments equal to any two similar arcs.

5. When a figure varies its magnitude and retains its form, any similar figure may be considered as one stage in its variation.

Hence the above relations, 1, 2, 3, 4, may be stated as follows:—

The area of any figure with constant form varies as the square upon any one of its line-segments.

Exercises.

1. Two triangles having one angle in each equal have their areas proportional to the rectangles on the sides containing the equal angles.
2. Two equal triangles, which have an angle in each equal, have the sides about this angle reciprocally proportional, *i.e.*, $a : a' = b' : b$.
3. The circle described on the hypothenuse of a right-angled triangle is equal to the sum of the circles described on the sides as diameters.
4. If semicircles be described outwards upon the sides of a right-angled triangle and a semicircle be described inwards on the hypothenuse, two crescents are formed whose sum is the area of the triangle.
5. AB is bisected in C, D is any point in AB, and the curves are semicircles. Prove that $P + S = Q + R$.
6. If a, b denote adjacent sides of a parallelogram and also of a rectangle, the ratio of the area of the parallelogram to that of the rectangle is the sine of the angle of the parallelogram.

7. The sides of a concyclic quadrangle are a, b, c, d. Then the cosine of the angle between a and b is
$$(a^2+b^2-c^2-d^2)/2(ab+cd).$$
8. In the quadrangle of 7, if s denotes one-half the perimeter,
$$\text{area}=\sqrt{\{(s-a)(s-b)(s-c)(s-d)\}}.$$
9. In any parallelogram the ratio of the rectangle on the sum and differences of adjacent sides to the rectangle on the diagonals is the cosine of the angle between the diagonals.
10. If a, b be the adjacent sides of a parallelogram and θ the angle between them, one diagonal is double the other when $\cos\theta = \dfrac{3}{10}\left(\dfrac{a}{b}+\dfrac{b}{a}\right)$.
11. If one diagonal of a parallelogram is expressed by $\sqrt{\left\{\dfrac{2(a^2+b^2)}{n^2+1}\right\}}$, the other diagonal is n times as long.
12. Construct an isosceles triangle in which the altitude is a mean proportional between the side and the base.
13. Three circles touch two lines and the middle circle touches each of the others. Prove that the radius of the middle circle is a mean proportional between the radii of the others.
14. In an equilateral triangle describe three circles which shall touch one another and each of which shall touch a side of the triangle.
15. In an equilateral triangle a circle is described to touch the incircle and two sides of the triangle. Show that its radius is one-third that of the incircle.

PART IV.

SECTION I.

GEOMETRIC EXTENSIONS.

220°. Let two lines L and M passing through the fixed points A and B meet at P.

When P moves in the direction of the arrow, L and M approach towards parallelism, and the angle APB diminishes. Since the lines are unlimited (21°, 3) P may recede from A along L until the segment AP becomes greater than any conceivable length, and the angle APB becomes less than any conceivable angle.

And as this process may be supposed to go on endlessly, P is said to "go to infinity" or to "be at infinity," and the ∠APB is said to vanish.

But lines which make no angle with one another are parallel, ∴ *Parallel lines meet at infinity, and lines which meet at infinity are parallel.*

The symbol for "infinity" is ∞.

The phrases "to go to infinity," "to be at infinity," must not be misunderstood. Infinity is not a place but a property. Lines which meet at ∞ are lines so situated that, having the same direction they cannot meet at any finite point, and therefore cannot meet at all, within our apprehension, since every point that can be conceived of is finite.

GEOMETRIC EXTENSIONS.

The convenience of the expressions will appear throughout the sequel.

Cor. Any two lines in the same plane meet: at a finite point if the lines are not parallel, at infinity if the lines are parallel.

221°. L and M are lines intersecting in O, and P is any point from which PB and PA are ∥ respectively to L and M. A third and variable line N turns about P in the direction of the arrow.

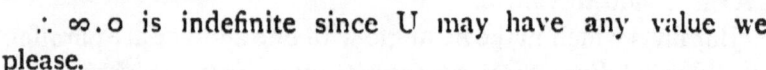

1. AX . BY = a constant (184°)
   = U say.

When N comes to parallelism with L, AX becomes infinite and BY becomes zero.

∴ ∞ . o is indefinite since U may have any value we please.

2. The motion continuing, let N come into the position N'. Then AX' is opposite in sense to AX, and BY' to BY. But AX increased to ∞, changed sign and then decreased absolutely, until it reached its present value AX', while BY decreased to zero and then changed sign.

∴ a magnitude changes sign when it passes through zero or infinity.

3. It is readily seen that, as the rotation continues, BY' increases negatively and AX' decreases, as represented in one of the stages of change at X" and Y". After this Y" goes off to ∞ as X" comes to A. Both magnitudes then change sign again, this time BY" by passing through ∞ and AX" by passing through zero.

Since both segments change sign together the product or rectangle remains always positive and always equal to the constant area U.

222°. A line in the plane admits of one kind of variation, rotation. When it rotates about a fixed finite point it describes angles about that point. But since all the lines of a system of parallels meet at the same point at infinity, rotation about that point is equivalent to translation, without rotation, in a direction orthogonal to that of the line.

Hence any line can be brought into coincidence with any other line in its plane by rotation about the point of intersection.

223°. If a line rotates about a finite point while the point simultaneously moves along the line, the point traces a curve to which the line is at all times a tangent. The line is then said to envelope the curve, and the curve is called the *envelope* of the line.

The algebraic equation which gives the relation between the rate of rotation of the line about the point and the rate of translation of the point along the line is the *intrinsic* equation to the curve.

224°. A line-segment in the plane admits of two kinds of variation, viz., variation in length, and rotation.

If one end-point be fixed the other describes some locus depending for its character upon the nature of the variations.

The algebraic equation which gives the relation between the rate of rotation and the rate of increase in length of the segment, or radius vector, is the *polar* equation of the locus.

When the segment is invariable in length the locus is a circle.

225°. A line which, by rotation, describes an angle may rotate in the direction of the hands of a clock or in the contrary direction.

If we call an angle described by one rotation positive we must call that described by the other negative. Unless convenience requires otherwise, the direction of rotation of the hands of a clock is taken as negative.

An angle is thus counted from zero to a circumangle either positively or negatively.

The angle between AB and A'B' is the rotation which brings AB to A'B', and is either $+\alpha$ or $-\beta$, and the sum of these two angles irrespective of sign is a circumangle.

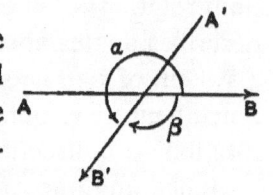

When an angle exceeds a circumangle the excess is taken in Geometry as the angle.

Ex. QA and QB bisect the angles CAB and ABP externally ; to prove that $\angle P = 2\angle Q$.

The rotation which brings CP to AB is -2α, AB to BP is $+2\beta$,

$$\therefore \angle P = 2(\beta - \alpha).$$

Also, the rotation which brings AQ to AB is $-\alpha$, and AB to BQ is $+\beta$, $\therefore \angle Q = \beta - \alpha$.

$$\therefore \qquad \angle(CP . BP) = 2\angle(AQ . BQ).$$

This property is employed in the working of the sextant.

226°. Let AB and CD be two diameters at right angles. The rectangular sections of the plane taken in order of positive rotation and starting from A are called respectively the first, second, third, and fourth quadrants, the first being AOC, the second COB, etc.

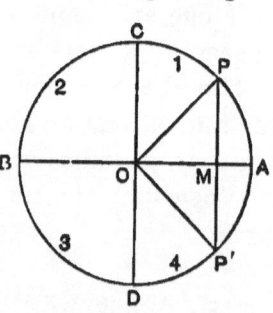

The radius vector starting from coincidence with OA may describe the positive $\angle AOP$, or the negative $\angle AOP'$.

Let these angles be equal in absolute value, so that the $\triangle MOP \equiv \triangle MOP'$, PM being $\perp$ on OA.

Then $PM = -P'M$, since in passing from P to P', PM passes through zero.

$$\therefore \qquad \sin AOP' = \frac{P'M}{OP} = \frac{-PM}{OP} = -\sin AOP.$$

and $\cos AOP' = \dfrac{OM}{OP'} = \dfrac{OM}{OP} = \cos AOP.$

∴ the sine of an angle changes sign when the angle does, but the cosine does not.

227°. As the angle AOP increases, OP passes through the several quadrants in succession.

When OP lies in the 1st Q., sin AOP and cos AOP are both positive; when OP lies in the 2nd Q., sin AOP is positive and cos AOP is negative; when OP lies in the 3rd Q., the sine and cosine are both negative; and, lastly, when OP lies in the 4th Q., sin AOP is negative and the cosine positive.

Again, when P is at A, ∠AOP=0, and PM=0, while OM=OP. ∴ sin 0=0 and cos 0=1.

When P comes to C, PM=OP and OM=0, and denoting a right angle by $\dfrac{\pi}{2}$, (207°, Cor. 4)

$\sin \dfrac{\pi}{2} = 1$, and $\cos \dfrac{\pi}{2} = 0.$

When P comes to B, PM=0 and OM=−1,

∴ $\sin \pi = 0$, and $\cos \pi = -1.$

Finally when P comes to D, PM=−OP and OM=0.

∴ $\sin \dfrac{3\pi}{2} = -1$, and $\cos \dfrac{3\pi}{2} = 0.$

These variations of the sine and cosine for the several quadrants are collected in the following table:—

	1st Q.		2nd Q.		3rd Q.		4th Q.	
Sine,	+		+		−		−	
Cosine,	+		−		−		+	
	From	To	From	To	From	To	From	To
Sine,	0	1	1	0	0	−1	−1	0
Cosine,	1	0	0	−1	−1	0	0	1

228°. ABC is a triangle in its circumcircle whose diameter we will denote by d.

Let CD be a diameter.

Then $\angle D = \angle A$, (106°, Cor. 1)
and $\angle CBD = \sqsupset$.

$\therefore \quad CB = CD \sin CDB = d \sin A = a$.

and from symmetry,

$$d = \frac{a}{\sin A} = \frac{b}{\sin B} = \frac{c}{\sin C}.$$

Hence the sides of a triangle are proportional to the sines of the opposite angles; and the diameter of the circumcircle is the quotient arising from dividing any side by the sine of the angle opposite that side.

PRINCIPLE OF ORTHOGONAL PROJECTION.

229°. The *orthogonal projection* (167°, 2) of PQ on L is P'Q', the segment intercepted between the feet of the perpendiculars PP' and QQ'.

Now $P'Q' = PQ \cos (PQ \, . \, P'Q')$.

$\therefore$ the projection of any segment on a given line is the segment multiplied by the cosine of the angle which it makes with the given line.

From left to right being considered as the + direction along L, the segment PQ lies in the 1st Q., as may readily be seen by considering P, the point *from* which we read the segment, as being the centre of a circle through Q.

Similarly QP lies in the 3rd Q., and hence the projection of PQ on L is + while that of QP is −.

When PQ is $\perp$ to L, its projection on L is zero, and when $\parallel$ to L this projection is PQ itself.

Results obtained through orthogonal projection are universally true for all angles, but the greatest care must be exercised with regard to the signs of angular functions concerned.

Ex. AX and OY are fixed lines at right angles, and AQ is any line and P any point.

Required to find the $\perp$PQ in terms of AX, PX, and the $\angle$A.

Take PQ as the positive direction, and project the closed figure PQAXP on the line of PQ. Then

pr.PQ + pr.QA + pr.AX + pr.XP = 0. (168°)

Now, pr.PQ is PQ, and pr.QA = 0; AX lies in 1st Q., and XP in the 3rd Q.

Moreover $\angle$(AX . PQ), *i.e.*, the rotation which brings AX to PQ in direction is $-\angle$N, and its cosine is $+$.

$\therefore$ $\qquad$ cos $\angle$(AX . PQ) = + sin A.

Also, pr.XP is $-$ XP cos $\angle$XPQ = $-$ XP cos A.

$\therefore$ $\qquad$ PQ = XP cos A $-$ AX sin A.

SIGNS OF THE SEGMENTS OF DIVIDED LINES AND ANGLES.

230°. AOB is a given angle and $\angle$AOB = $-\angle$BOA.

Let OP divide the $\angle$AOB internally, and OQ divide it externally into parts denoted respectively by a, β, and $a'\beta'$.

If a is the $\angle$AOP and β the $\angle$POB, a and β are both positive. But if we write Q for P, $a' = \angle$AOQ, and $\beta' = \angle$QOB, and a' and β' have contrary signs.

On the other hand, if a is $\angle$AOP and β the $\angle$BOP, a and β have contrary signs, while replacing P by Q gives a' and β' with like signs.

The choice between these usages must depend upon convenience; and as it is more symmetrical with a two-letter notation to write AOP, BOP, AOQ, BOQ, than AOP, POB, etc., we adopt the convention that internal division of an angle gives segments with opposite signs, while external division gives segments with like signs.

In like manner the internal division of the segment AB gives parts AP, BP having unlike signs, while external division gives parts AQ, BQ having like signs.

Def.—A set of points on a line is called a *range*, and the line is called its *axis*.

By connecting the points of the range with any point not on its axis we obtain a corresponding pencil.   (203°, Def.)

Cor. To any range corresponds a pencil for every vertex, and to any vertex corresponds a range for every axis, the axis being a transversal to the rays of the pencil.

If the vertex is on the axis the rays are coincident; and if the axis passes through the vertex the points are coincident.

231°. BY is any line dividing the angle B, and CR, AP are perpendiculars upon BY.

Then $\triangle APY \backsim \triangle CRY$,
and   AP is AB sin ABY,
and   CR is BC sin CBY,
∴ $\dfrac{AY}{CY} = \dfrac{AB}{CB} \cdot \dfrac{\sin ABY}{\sin CBY}$.

Therefore a line through the vertex of a triangle divides the base into segments which are proportional to the products of each conterminous side multiplied by the sine of the corresponding segment of the vertical angle.

Cor. 1. Let BY bisect ∠B, then $\dfrac{AY}{YC} = \dfrac{c}{a}$.

∴ $AY = \dfrac{c}{a}(b - AY)$, and $AY = \dfrac{bc}{a+c}$.

Thence $YC = \dfrac{ba}{a+c}$.

Which are the segments into which the bisector of the $\angle B$ divides the base AC.

Cor. 2. In the $\triangle ABY$,
$$BY^2 = AB^2 + AY^2 - 2AB \cdot AY \cdot \cos A. \qquad (217°)$$

But $\cos A = \dfrac{b^2 + c^2 - a^2}{2bc}$ (217°), and $AY = \dfrac{bc}{a+c}$,

whence by reduction
$$BY^2 = ac\left\{ 1 - \left(\dfrac{b}{a+c}\right)^2 \right\},$$

which is the square of the length of the bisector.

Cor. 3. When $AY = CY$, BY is a median, and
$$\dfrac{AB}{CB} = \dfrac{\sin YBC}{\sin ABY}.$$

∴ a median to a triangle divides the angle through which it passes into parts whose sines are reciprocally as the conterminous sides.

232°. In any range, when we consider both sign and magnitude, the sum $AB + BC + CD + DE + EA = 0$, however the points may be arranged.

For, since we start from A and return to A, the translation in a + direction must be equal to that in a − direction.

That this holds for any number of points is readily seen.

Also, in any pencil, when we consider both sign and magnitude, the sum $\angle AOB + \angle BOC + \angle COD + \angle DOA = 0$.

For we start from the ray OA and end with the ray OA, and hence the rotation in a + direction is equal to that in a − direction.

RANGES AND PENCILS OF FOUR.

233°. Let A, B, C, P be a range of four, then
$$AB \cdot CP + BC \cdot AP + CA \cdot BP = 0.$$

GEOMETRIC EXTENSIONS. 187

Proof.— $AP = AC + CP$, and $BP = BC + CP$.
∴ the expression becomes
$$BC(AC + CA) + (AB + BC + CA)CP,$$
and each of the brackets is zero (232°). ∴ etc.

234°. Let $O.ABCP$ be a pencil of four. Then
$\sin AOB . \sin COP + \sin BOC . \sin AOP + \sin COA . \sin BOP = 0$.

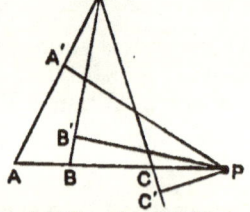

Proof.—$\triangle AOB = \tfrac{1}{2} OA . OB \sin AOB$,
also $\triangle AOB = \tfrac{1}{2} AB . p$,
where p is the common altitude to all the triangles.
∴ $AB . p = OA . OB . \sin AOB$.
Similarly, $CP . p = OC . OP . \sin COP$.
∴ $AB . CP . p^2 = OA . OB . OC . OP \sin AOB . \sin COP$.
Now, p^2 and $OA . OB . OC . OP$ appear in every homologous product, ∴ $(AB . CP + BC . AP + CA . BP) p^2$
$= OA . OB . OC . OP (\sin AOB . \sin COP$
$\qquad + \sin BOC . \sin AOP + \sin COA . \sin BOP)$.
But the bracket on the left is zero (233°), and $OA.OB.OC.OD$ is not zero, therefore the bracket on the right is zero. *q.e.d.*

235°. From P let perpendiculars PA', PB', PC' be drawn to OA, OB, and OC respectively. Then
$$\sin AOP = \frac{A'P}{OP}, \quad \sin BOP = \frac{B'P}{OP}, \quad \text{etc.,}$$
and putting these values for $\sin AOP$, etc., in the relation of 234°, we have, after multiplying through by OP,
$$C'P . \sin AOB + A'P . \sin BOC + B'P . \sin COA = 0.$$
Or, let L, M, and N be any three concurrent lines, l, m, n the perpendiculars from any point P upon L, M, and N respectively, then
$$l \sin \widehat{MN} + m \sin \widehat{NL} + n \sin \widehat{LM} = 0.$$
where $\widehat{MN}$ denotes the angle between M and N, etc.

236°. Ex. 1. Let four rays be disposed in the order OA, OB, OC, OP, and let OP be perpendicular to OA.

Denote $\angle AOC$ by A, and $\angle AOB$ by B. Then 234° becomes
$$\sin B \cos A + \sin(A-B)\sin\tfrac{\pi}{2} - \sin A \cos B = 0,$$
or, $\sin(A-B) = \sin A \cos B - \cos A \sin B$.

Similarly, by writing the rays in the same order and making $\angle BOP$ a $\sqsupset$, and denoting $\angle AOB$ by A and $\angle BOC$ by B, we obtain $\sin(A+B) = \sin A \cos B + \cos A \sin B$.

Also, by writing the rays in the order OA, OP, OB, OC, and denoting $\angle AOP$ by A and $\angle BOC$ by B, we obtain

(1) when $\angle AOB = \sqsupset$,
$\cos(A-B) = \cos A \cos B + \sin A \sin B$;

(2) when $\angle AOC = \sqsupset$,
$\cos(A+B) = \cos A \cos B - \sin A \sin B$;

which are the addition theorems for the sine and cosine.

Ex. 2. ABC is a triangle and P is any point. Let PX, PY, PZ be perpendiculars upon BC, CA, AB, and be denoted by P_a, P_b, P_c respectively.

Draw AQ $\parallel$ to BC to meet PX in Q. Then (235°)
$$PQ \sin A + PY \sin B + PZ \sin C = 0.$$
But if AD is $\perp$ to BC, AD $= b \sin C = $ QX.

$\therefore \quad (PX - b \sin C)\sin A + PY \sin B + PZ \sin C = 0,$

or $\quad \Sigma(P_a \sin A) = b \sin A \sin C.$

Similarly, $\quad \Sigma(P_a \sin A) = c \sin B \sin A$
$\qquad\qquad = a \sin C \sin B,$

$\therefore \quad \Sigma(P_a \sin A) = \sqrt[3]{\{abc \sin^2 A \sin^2 B \sin^2 C\}}.$

Hence the function of the perpendicular
$$P_a \sin A + P_b \sin B + P_c \sin C$$
is constant for all positions of P. This constancy is an important element in the theory of *trilinear co-ordinates*.

237°. A, B, C being a range of three, and P any point not on the axis,
$$AB \cdot CP^2 + BC \cdot AP^2 + CA \cdot BP^2$$
$$= - AB \cdot BC \cdot CA.$$

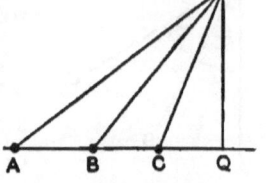

Proof.—Let PQ be ⊥ to AC. Then
$$AQ = AC + CQ, \quad BQ = BC + CQ,$$
and the expression becomes
$$(AB + BC + CA)(PQ^2 + CQ^2) + BC \cdot CA(BC - AC)$$
$$\equiv BC \cdot CA(BC + CA) \equiv BC \cdot CA \cdot BA$$
$$\equiv - AB \cdot BC \cdot CA.$$

EXERCISES.

1. A number of stretched threads have their lower ends fixed to points lying in line on a table, and their other ends brought together at a point above the table. What is the character of the system of shadows on the table when (*a*) a point of light is placed at the same height above the table as the point of concurrence of the threads? (*b*) when placed at a greater or less height?

2. If a line rotates uniformly about a point while the point moves uniformly along the line, the point traces and the line envelopes a circle.

3. If a radius vector rotates uniformly and at the same time lengthens uniformly, obtain an idea of the curve traced by the distal end-point.

4. Divide an angle into two parts whose sines shall be in a given ratio. (Use 231°, Cor. 3.)

5. From a given angle cut off a part whose sine shall be to that of the whole angle in a given ratio.

6. Divide a given angle into two parts such that the product of their sines may be a given quantity. Under what condition is the solution impossible?

7. Write the following in their simplest form:—
$$\sin(\pi - \theta), \quad \sin\left(\tfrac{\pi}{2} + \theta\right), \quad \sin\left\{\pi - \left(\theta - \tfrac{\pi}{2}\right)\right\},$$
$$\cos(2\pi + \theta), \quad \cos\left\{2\pi - \left(\theta - \tfrac{\pi}{2}\right)\right\}, \quad \cos\left\{\theta - \left(\tfrac{\pi}{2} + \theta\right)\right\}.$$

8. Make a table of the variation of the tangent of an angle in magnitude and sign.
9. OM and ON are two lines making the $\angle MON = \omega$, and PM and PN are perpendiculars upon OM and ON respectively. Then $OP \sin \omega = MN$.
10. A transversal makes angles A', B', C' with the sides BC, CA, AB of a triangle. Then
$$\sin A \sin A' + \sin B \sin B' + \sin C \sin C' = 0.$$
11. OA, OB, OC, OP being four rays of any length whatever,
$$\triangle AOB \cdot \triangle COP + \triangle BOC \cdot \triangle AOP + \triangle COA \cdot \triangle BOP = 0.$$
12. If r be the radius of the incircle of a triangle, and r_1 be that of the excircle to side a, and if p_1 be the altitude to the side a, etc.,
$$p_1 = \frac{r}{\sin A}(\sin A + \sin B + \sin C)$$
$$= \frac{r_1}{\sin A}(-\sin A + \sin B + \sin C),$$
and $\quad \dfrac{1}{r_1} + \dfrac{1}{r_2} + \dfrac{1}{r_3} = \dfrac{1}{p_1} + \dfrac{1}{p_2} + \dfrac{1}{p_3} = \dfrac{1}{r}.$ (Use 235°.)
13. The base AC of a triangle is trisected at M and N, then
$$BN^2 = \tfrac{1}{9}(3BC^2 + 6BA^2 - 2AC^2).$$

SECTION II.

CENTRE OF MEAN POSITION.

238°. A, B, C, D are any points in line, and perpendiculars AA', BB', etc., are drawn to any fixed line L. Then there is, on the line, evidently some point, O, for which
$$AA' + BB' + CC' + DD' = 4ON\ ;$$
and ON is less than AA' and greater than DD'.

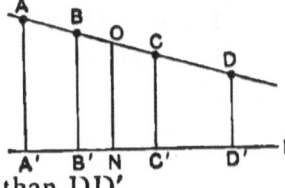

The point O is called the *centre of mean position*, or simply the *mean centre*, of the system of points A, B, C, D.

CENTRE OF MEAN POSITION.

Again, if we take multiples of the perpendiculars, as $a.AA'$, $b.BB'$, etc., there is some point O, on the axis of the points, for which
$$a.AA'+b.BB'+c.CC'+d.DD'=(a+b+c+d)ON.$$
Here again ON lies between AA' and DD'.

O is then called the *mean centre* of the system of points for the system of multiples.

Def.— For a range of points with a system of multiples we define the mean centre by the equation
$$\Sigma(a.AO)=0,$$
where $\Sigma(a.AO)$ is a contraction for
$$a.AO+b.BO+c.CO+...,$$
and the signs and magnitudes of the segments are both considered.

The notion of the mean centre or centre of mean position has been introduced into Geometry from Statics, since a system of material points having their weights denoted by a, b, c, ..., and placed at A, B, C, ..., would "balance" about the mean centre O, if free to rotate about O under the action of gravity.

The mean centre has therefore a close relation to the "centre of gravity" or "mass centre" of Statics.

239°. *Theorem.*—If P is an independent point in the line of any range, and O is the mean centre,
$$\Sigma(a.AP)=\Sigma(a).OP.$$

Proof.—AP = AO + OP,   BP = BO + OP, etc.,
∴ $a.AP = a.AO + a.OP$, $b.BP = b.BO + b.OP$, etc.
. . . .
∴ $\Sigma(a.AP) = \Sigma(a.AO) + \Sigma(a).OP.$
But, if O is the mean centre,
    $\Sigma(a.AO) = 0$, by definition,
∴ $\Sigma(a.AP) = \Sigma(a).OP.$

Ex. The mean centre of the basal vertices of a triangle

when the multiples are proportional to the opposite sides is the foot of the bisector of the vertical angle.

240°. Let A, B, C, ... be a system of points situated anywhere in the plane, and let AL, BL, CL, ..., AM, BM, CM, ..., denote perpendiculars from A, B, C, ... upon two lines L and M.

Then we define the mean centre of the system of points for a system of multiples as the point of intersection of L and M

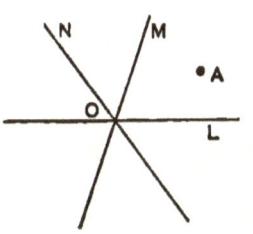

when $\Sigma(a \cdot AL) = 0$,
and $\Sigma(a \cdot AM) = 0$.

If N be any other line through this centre, $\Sigma(a \cdot AN) = 0$.

For, let A be one of the points. Then, since L, M, N is a pencil of three and A any point, (235°)

$AL \cdot \sin MON + AM \cdot \sin NOL + AN \cdot \sin LOM = 0$,

also $BL \cdot \sin MON + BM \cdot \sin NOL + BN \cdot \sin LOM = 0$,

. . . .

and multiplying the first by a, the second by b, etc., and adding,
$\Sigma(a \cdot AL)\sin MON + \Sigma(a \cdot AM)\sin NOL + \Sigma(a \cdot AN)\sin LOM = 0$.
But $\Sigma(a \cdot AL) = \Sigma(a \cdot AM) = 0$, by definition,
∴ $\Sigma(a \cdot AN) = 0$.

241°. *Theorem.*—If O be the mean centre of a system of points for a system of multiples, and L any line whatever,
$$\Sigma(a \cdot AL) = \Sigma(a) \cdot OL.$$

Proof.—Let M be ∥ to L and pass through O. Then
AL = AM + ML, ∴ $a \cdot AL = a \cdot AM + a \cdot ML$,
BL = BM + ML, ∴ $b \cdot BL = b \cdot BM + b \cdot ML$,
. . . .
adding, $\Sigma(a \cdot AL) = \Sigma(a \cdot AM) + \Sigma(a) \cdot ML$.
But, since M passes through O,
$\Sigma(a \cdot AM) = 0$, and ML = OL,
∴ $\Sigma(a \cdot AL) = \Sigma(a) \cdot OL$.

242°. Theorem.—The mean centre of the vertices of a triangle with multiples proportional to the opposite sides is the centre of the incircle.

Proof.—Take L along one of the sides, as BC, and let p be the $\perp$ from A. Then
$$\Sigma(a \cdot \mathrm{AL}) = a \cdot p$$
and $\qquad \Sigma(a) \cdot \mathrm{OL} = (a+b+c) \cdot \mathrm{OL},$

$\therefore$ (239°) $\qquad \mathrm{OL} = \dfrac{ap}{a+b+c} = \dfrac{\triangle}{s} = r;$ (153°, Ex. 1)

i.e., the mean centre is at the distance r from each side, and is the centre of the incircle.

Cor. 1. If one of the multiples, as a, be taken negative,
$$\mathrm{OL} = \dfrac{-ap}{-a+b+c} = \dfrac{-\triangle}{s-a} = -r'; \qquad (153°, \text{Ex. 2})$$

i.e., the mean centre is beyond L, and is at the distance r' from each side, or it is the centre of the excircle to the side a.

Cor. 2. If any line be drawn through the centre of the incircle of a triangle, and α, β, γ be the perpendiculars from the vertices upon it, $\qquad a\alpha + b\beta + c\gamma = 0,$
and if the line passes through the centre of an excircle, that on the side a for example, $\quad a\alpha = b\beta + c\gamma.$

EXERCISES.

1. If a line so moves that the sum of fixed multiples of the perpendiculars upon it from any number of points is constant, the line envelopes a circle whose radius is
$$r = \dfrac{\Sigma(a \cdot \mathrm{AL})}{\Sigma(a)}.$$

2. The mean centre of the vertices of a triangle, for equal multiples, is the centroid.

3. The mean centre of the vertices of any regular polygon, for equal multiples, is the centre of its circumcircle.

243°. Theorem.—If O be the mean centre of a system of points for a system of multiples, and P any independent point in the plane,

$$\Sigma(a \cdot AP^2) = \Sigma(a \cdot AO^2) + \Sigma(a) \cdot OP^2.$$

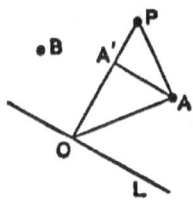

Proof.—Let O be the mean centre, P the independent point, and A any point of the system. Let L pass through O and be perpendicular to OP, and let AA' be perpendicular to OP. Then

$$AP^2 = AO^2 + OP^2 - 2OP \cdot OA',$$

and $\quad a \cdot AP^2 = a \cdot AO^2 + a \cdot OP^2 - 2OP \cdot a \cdot OA'.$
Similarly $\quad b \cdot BP^2 = b \cdot BO^2 + b \cdot OP^2 - 2OP \cdot b \cdot OB',$

. . . .

$\therefore \quad \Sigma(a \cdot AP^2) = \Sigma(a \cdot AO^2) + \Sigma(a) \cdot OP^2 - 2OP \cdot \Sigma(a \cdot OA').$
But $\quad \Sigma(a \cdot OA') = \Sigma(a \cdot AL) = 0, \quad\quad (241°)$
$\therefore \quad \Sigma(a \cdot AP^2) = \Sigma(a \cdot AO^2) + \Sigma(a) \cdot OP^2. \quad\quad q.e.d.$

Cor. In any regular polygon of n sides $\dfrac{1}{n}$th the sum of the squares on the joins of any point with the vertices is greater than the square on the join of the point with the mean centre of the polygon by the square on the circumradius.

For making the multiples all unity,

$$\Sigma(AP^2) = nr^2 + nOP^2,$$

$\therefore \quad \dfrac{1}{n}\Sigma(AP^2) = OP^2 + r^2.$

Ex. Let a, b, c be the sides of a triangle, and α, β, γ the joins of the vertices with the centroid. Then (242°, Ex. 2)

$$\Sigma(AP^2) = \Sigma(AO^2) + 3OP^2.$$

1st. Let P be at A, $\quad b^2 + c^2 = \alpha^2 + \beta^2 + \gamma^2 + 3\alpha^2,$
2nd. ,, P ,, B, $\quad c^2 + a^2 = \alpha^2 + \beta^2 + \gamma^2 + 3\beta^2,$
3rd. ,, P ,, C, $\quad a^2 + b^2 = \alpha^2 + \beta^2 + \gamma^2 + 3\gamma^2,$
whence $\quad a^2 + b^2 + c^2 = 3(\alpha^2 + \beta^2 + \gamma^2).$

Ex. If ABCDEFGH be the vertices of a regular octagon taken in order, $AC^2 + AD^2 + AE^2 + AF^2 + AG^2 = 14r^2.$

244°. Let O be the centre of the incircle of the $\triangle ABC$ and let P coincide with A, B, and C in succession.

1st. $\qquad bc^2 + cb^2 = \Sigma(a \cdot AO^2) + \Sigma(a)AO^2,$
2nd. $\qquad ac^2 + ca^2 = \Sigma(a \cdot AO^2) + \Sigma(a)BO^2,$
3rd. $\qquad ab^2 + ba^2 = \Sigma(a \cdot AO^2) + \Sigma(a)CO^2.$

Now, multiply the 1st by a, the 2nd by b, the 3rd by c, and add, and we obtain, after dividing by $(a+b+c)$,
$$\Sigma(a \cdot AO^2) = abc.$$

Cor. 1. For any triangle, with O as the centre of the incircle, the relation $\Sigma(a \cdot AP^2) = \Sigma(a \cdot AO^2) + \Sigma(a)OP^2$
becomes $\qquad \Sigma(a \cdot AP^2) = abc + 2s \cdot OP^2,$
and, if O be the centre of an excircle on side a, for example,
$$\Sigma(\bar{a} \cdot AP^2) = -abc + 2(s-a)OP^2,$$
where $\bar{a}$ denotes that a alone is negative.

Cor. 2. Let P be taken at the circumcentre, and let D be the distance between the circumcentre and the centre of the incircle. Then $\qquad AP = BP = CP = R.$
$\therefore \qquad 2sR^2 = abc + 2sD^2.$
But $\qquad abc = 4\triangle R, \qquad\qquad$ (204°, Cor.)
and $\qquad s = \dfrac{\triangle}{r}, \qquad\qquad$ (153°, Ex. 1)
$\therefore \qquad D^2 = R^2 - 2Rr.$

Cor. 3. If D_1 be the distance between the circumcentre and the centre of an excircle to the side a, we obtain in a similar manner $\qquad D_1^2 = R^2 + 2Rr_1.$
Similarly $\qquad D_2^2 = R^2 + 2Rr_2,$
$\qquad\qquad D_3^2 = R^2 + 2Rr_3.$

245°. Ex. To find the product $OA \cdot OB \cdot OC$, where O is the centre of the incircle.

Let P coincide with A. Then (244°)
$$bc^2 + b^2c = abc + 2s \cdot AO^2,$$
$\therefore \qquad AO^2 = \dfrac{bc(s-a)}{s}.$

Similarly $\quad BO^2 = \dfrac{ca(s-b)}{s}$, and $CO^2 = \dfrac{ab(s-c)}{s}$,

∴ $\quad AO^2 \cdot BO^2 \cdot CO^2 = \dfrac{a^2b^2c^2 s(s-a)(s-b)(s-c)}{s^4}$

$$= \dfrac{a^2b^2c^2}{s^4}\Delta^2 = 16R^2r^4,$$

and $\quad OA \cdot OB \cdot OC = 4Rr^2$.

246°. If $\Sigma(a \cdot AP^2)$ becomes constant, k, we have
$$k = \Sigma(a \cdot AO^2) + \Sigma(a)OP^2,$$
and $\Sigma(a \cdot AO^2)$ being independent of the position of P, and therefore constant for variations of P, OP is also constant, and P describes a circle whose radius is
$$R^2 = \dfrac{k - \Sigma(a \cdot AO^2)}{\Sigma(a)}.$$

∴ If a point so moves that the sum of the squares of its joins with any number of fixed points, each multiplied by a given quantity, is constant, the point describes a circle whose centre is the mean point of the system for the given multiples.

Exercises.

1. If O', O'', O''' be the centres of the escribed circles,
$$AO' \cdot BO'' \cdot CO''' = 4Rs^2.$$
2. $\quad AO' \cdot BO' \cdot CO' = 4Rr_1^2.$
3. $\quad s \cdot OL = (s-a)O'L + (s-b)O''L + (s-c)O'''L,$
where L is any line whatever.
4. If P be any point,
$$s \cdot OP^2 = (s-a)O'P^2 + (s-b)O''P^2 + (s-c)O'''P^2 - 2abc.$$
5. $\quad (s-a) \cdot O'O^2 + (s-b) \cdot O''O^2 + (s-c)O'''O^2 = 2abc.$
6. If D is the distance between the circumcentre and the centroid, $\quad D^2 = \tfrac{1}{9}(9R^2 - a^2 - b^2 - c^2).$

SECTION III.

OF COLLINEARITY AND CONCURRENCE.

247°. *Def.* 1.—Three or more points in line are collinear, and three or more lines meeting in a point are concurrent.

Def. 2.—A *tetragram* or general quadrangle is the figure formed by four lines no three of which are concurrent, and no two of which are parallel.

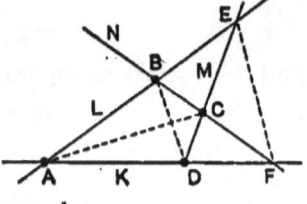

Thus L, M, N, K form a tetragram. A, B, C, D, E, F are its six vertices. AC, BD are its *internal* diagonals, and EF is its *external* diagonal.

248°. The following are promiscuous examples of collinearity and concurrence.

Ex. 1. AC is a ▱, and P is any point. Through P, GH is drawn ∥ to BC, and EF ∥ to AB.

The diagonals EG, HF, and DB of the three ▱s AP, PC, and AC are concurrent.

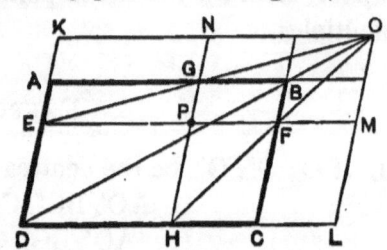

EG and HF meet in some point O; join BO and complete the ▱ OKDL, and make the extensions as in the figure.

We are to prove that D, B, O are collinear.

Proof.—▱KG = ▱GM, and ▱FL = ▱FN, (145°)
∴   ▱KG = ▱GF + ▱BM,
and   ▱FL = ▱GF + ▱NB.
Hence   ▱KB = ▱BL,
∴   B is on the line DO,   (145°, Cor. 2)
and   D, B, O are collinear.

Ex. 2. The middle points of the three diagonals of a tetragram are collinear.

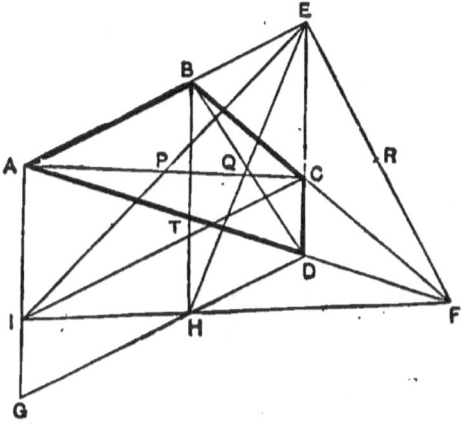

ABCDEF is the tetragram, P, Q, R the middle points of the diagonals.

Complete the parallelogram AEDG, and through B and C draw lines ∥ to AG and AE respectively, and let them meet in T.

Then, from Ex. 1, IH passes through F. Therefore EIHF is a triangle, and the middle points of EI, EH, and EF are collinear.

(84°, Cor. 2)

But these are the middle points of AC, BD, and EF respectively. ∴ P, Q, R are collinear.

Ex. 3. *Theorem.*—The circumcentre, the centroid, and the orthocentre of a triangle are collinear.

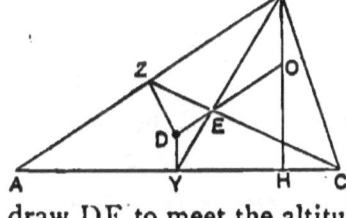

Proof.—Let YD and ZD be the right bisectors of AC and AB. Join BY, CZ, and through E, the intersection of these joins, draw DE to meet the altitude BH in O.

Then D is the circumcentre and E is the centroid. Since DY is ∥ to BH, the triangles YDE and BOE are similar.
But $\qquad$ BE = 2EY, $\qquad$ (85°, Cor.)
∴ $\qquad$ OE = 2DE,
and as D and E are fixed points, O is a fixed point.
∴ the remaining altitudes pass through O.

249°. *Theorem.*—Three concurrent lines perpendicular to the sides of a triangle at X, Y, Z divide the sides so that
$$BX^2 + CY^2 + AZ^2 = CX^2 + AY^2 + BZ^2;$$

and, conversely, if three lines perpendicular to the sides of a triangle divide the sides in this manner, the lines are concurrent.

Proof.—Let OX, OY, OZ be the lines.
Then $BX^2 - CX^2 = BO^2 - CO^2$, (172°, 1)
Similarly $CY^2 - AY^2 = CO^2 - AO^2$,
$AZ^2 - BZ^2 = AO^2 - BO^2$,
$\therefore\quad BX^2 + CY^2 + AZ^2 - CX^2 - AY^2 - BZ^2 = 0.$ *q.e.d.*

Conversely, let X, Y, Z divide the sides of the triangle in the manner stated, and let OX, OY, perpendiculars to BC and CA, meet at O. Then OZ is $\perp$ to AB.

Proof.—If possible let OZ' be $\perp$ to AB, Then, by the theorem, $BX^2 + CY^2 + AZ'^2 - CX^2 - AY^2 - BZ'^2 = 0$,
and by hyp. $BX^2 + CY^2 + AZ^2 - CX^2 - AY^2 - BZ^2 = 0$,
$\therefore\quad AZ'^2 - AZ^2 = BZ'^2 - BZ^2.$

But these differences have opposite signs and cannot be equal unless each is zero. $\therefore$ Z' coincides with Z.

Exercises.

1. When three circles intersect two and two, the common chords are concurrent.

    Let S, S_1, S_2 be the circles, and A, B, C their centres.
    Then (113°) the chords are perpendicular to the sides of the $\triangle ABC$ at X, Y, and Z. And if r, r_1, r_2 be the radii of the circles,
    $$BX^2 - CX^2 = r_1^2 - r_2^2, \text{ etc., etc.,}$$
    and the criterion is satisfied.
    $\therefore$ the chords are concurrent.

2. The perpendiculars to the sides of a triangle at the points of contact of the escribed circles are concurrent.

3. When three circles touch two and two the three common tangents are concurrent.

4. If perpendiculars from the vertices of one triangle on the sides of another be concurrent, then the perpendiculars

from the vertices of the second triangle on the sides of the first are concurrent.

5. When three perpendiculars to the sides of a triangle are concurrent, the other three at the same distances from the middle points of the sides are concurrent.

6. Two perpendiculars at points of contact of excircles are concurrent with a perpendicular at a point of contact of the incircle.

250°. *Theorem.*—When three points X, Y, Z lying on the sides BC, CA, and AB of a triangle are collinear, they divide the sides into parts which fulfil the relation

$$(a) \qquad \frac{BX \cdot CY \cdot AZ}{CX \cdot AY \cdot BZ} = 1;$$

and their joins with the opposite vertices divide the angles into parts which fulfil the relation

$$(b) \qquad \frac{\sin BAX \cdot \sin CBY \cdot \sin ACZ}{\sin CAX \cdot \sin ABY \cdot \sin BCZ} = 1.$$

Proof of (a).—On the axis of X, Y, Z draw the perpendiculars AP, BQ, CR.

On account of similar △s,

$$\frac{BX}{CX} = \frac{BQ}{CR}, \quad \frac{CY}{AY} = \frac{CR}{AP}, \quad \frac{AZ}{BZ} = \frac{AP}{BQ},$$

$$\therefore \frac{BX \cdot CY \cdot AZ}{CX \cdot AY \cdot BZ} = 1.$$

Proof of (b).—
$$\frac{BX}{CX} = \frac{\triangle BAX}{\triangle CAX} = \frac{BA \sin BAX}{CA \sin CAX},$$

$$\therefore \frac{\sin BAX}{\sin CAX} = \frac{CA}{BA} \cdot \frac{BX}{CX}.$$

Similarly, $\dfrac{\sin CBY}{\sin ABY} = \dfrac{AB}{CB} \cdot \dfrac{CY}{AY}, \quad \dfrac{\sin ACZ}{\sin BCZ} = \dfrac{BC}{AC} \cdot \dfrac{AZ}{BZ}.$

$$\therefore \frac{\sin BAX \cdot \sin CBY \cdot \sin ACZ}{\sin CAX \cdot \sin ABY \cdot \sin BCZ} = \frac{BX \cdot CY \cdot AZ}{CX \cdot AY \cdot BZ} = 1. \quad q.e.d.$$

The preceding functions which are criteria of collinearity

will be denoted by the symbols

$$\left(\frac{BX}{CX}\right) \text{ and } \left(\frac{\sin BAX}{\sin CAX}\right) \text{ respectively.}$$

It is readily seen that three points on the sides of a triangle can be collinear only when an even number of sides or angles (2 or 0) are divided internally, and from 230° it is evident that the sign of the product is + in these two cases.

Hence, in applying these criteria, the signs may be disregarded, as the final sign of the product is determined by the number of sides or angles divided internally.

The converses of these criteria are readily proved, and the proofs are left as an exercise to the reader.

Ex. If perpendiculars be drawn to the sides of a triangle from any point in its circumcircle, the feet of the perpendiculars are collinear.

X, Y, Z are the feet of the perpendiculars. If X falls between B and C, $\angle OBC$ is $<$ a ⌐, and therefore $\angle OAC$ is $>$ a ⌐, and Y divides AC externally;

∴ it is a case of collinearity.

Now, $\quad BX = OB \cos OBC,$
and $\quad AY = OA \cos OAC.$

But, neglecting sign, $\cos OBC = \cos OAC$,

∴ $\quad \dfrac{BX}{AY} = \dfrac{OB}{OA},$

and similarly, $\quad \dfrac{CY}{BZ} = \dfrac{OC}{OB}, \quad \dfrac{AZ}{CX} = \dfrac{OA}{OC}.$

∴ $\quad \left(\dfrac{BX}{CX}\right) = 1,$ and X, Y, Z are collinear.

Def.—The line of collinearity of X, Y, Z is known as "Simson's line for the point O."

251°. Theorem.—When three lines through the vertices of a triangle are concurrent, they divide the angles into parts which fulfil the relation

(a) $$\frac{\sin BAX \cdot \sin CBY \cdot \sin ACZ}{\sin CAX \cdot \sin ABY \cdot \sin BCZ} = -1,$$

and they divide the opposite sides into parts which fulfil the relation

(b) $$\frac{BX \cdot CY \cdot AZ}{CX \cdot AY \cdot BZ} = -1.$$

To prove (a).—Let O be the point of concurrence of AX, BY, and CZ, and let OP, OQ, OR be perpendiculars on the sides.

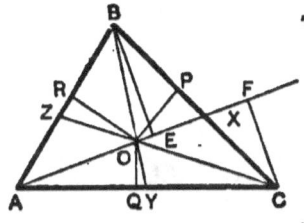

Then
$$\frac{\sin BAX}{\sin CAX} = \frac{OR}{OQ},$$
$$\frac{\sin CBY}{\sin ABY} = \frac{OP}{OR},$$
$$\frac{\sin ACZ}{\sin BCZ} = \frac{OQ}{OP},$$

∴ multiplying, $\left(\dfrac{\sin BAX}{\sin CAX}\right) = -1,$

the negative sign resulting from the three angles being divided internally. (230°)

To prove (b).—From B and C let BE and CF be perpendiculars upon AX.

Then, from similar △s BEX and CFX,
$$\frac{BX}{CX} = \frac{BE}{CF} = \frac{AB \sin BAX}{AC \sin CAX}.$$

Similarly, $\dfrac{CY}{AY} = \dfrac{BC \sin CBY}{BA \sin ABY}, \quad \dfrac{AZ}{BZ} = \dfrac{CA \sin ACZ}{CB \sin BCZ},$

∴ multiplying, $\left(\dfrac{BX}{CX}\right) = -1,$ from (a). *q.e.d.*

The negative sign results from the three sides being divided internally. (230°)

It is readily seen that three concurrent lines through the vertices of a triangle must divide an odd number of angles and of sides internally, and that the resulting sign of the product is accordingly negative.

Hence, in applying the criteria, the signs of the ratios may be neglected.

OF COLLINEARITY AND CONCURRENCE.

The remarkable relation existing between the criteria for collinearity of points and concurrence of lines will receive an explanation under the subject of Reciprocal Polars.

EXERCISES.

252°. 1. Equilateral triangles ABC', BCA', CAB' are described upon the sides AB, BC, CA of any triangle. Then the joins AA', BB', CC' are concurrent.

Proof.—Since $AC' = AB$, $AB' = AC$,
and $\angle CAC' = \angle BAB'$,
∴ $\triangle CAC' \equiv \triangle B'AB$, and $\angle AC'C = \angle ABB'$.

But $\dfrac{\sin ACZ}{\sin ABY} = \dfrac{\sin ACC'}{\sin AC'C} = \dfrac{AC'}{AC} = \dfrac{AB}{AC}$. (228°)

Similarly, $\dfrac{\sin BAX}{\sin BCZ} = \dfrac{BC}{BA}$, $\dfrac{\sin CBY}{\sin CAX} = \dfrac{CA}{CB}$

∴ $\left(\dfrac{\sin BAX}{\sin CAX}\right) = -1$,

and hence the joins AA', BB', CC' are concurrent.

2. The joins of the vertices of a △ with the points of contact of the incircle are concurrent.
3. The joins of the vertices of a △ with the points of contact of an escribed ⊙ are concurrent.
4. ABC is a △, right-angled at B, CD is = and ⊥ to CB, and AE is = and ⊥ to AB. Then EC and AD intersect on the altitude from B.
5. The internal bisectors of two angles of a △ and the external bisector of the third angle intersect the opposite sides collinearly.
6. The external bisectors of the angles of a △ intersect the opposite sides collinearly.
7. The tangents to the circumcircle of a △, at the vertices of the △, intersect the opposite sides collinearly.
8. If any point be joined to the vertices of a △, the lines through the point perpendicular to those joins intersect the opposite sides of the △ collinearly.

9. A ⊙ cuts the sides of a △ in six points so that three of them connect with the opposite vertices concurrently. Show that the remaining three connect concurrently with the opposite vertices.

10. Is the statement of Ex. 1 true when the △s are all described internally upon the sides of the given △?

11. If L is an axis of symmetry to the congruent △s ABC and A'B'C', and O is any point on L, A'O, B'O, and C'O intersect the sides BC, CA, and AB collinearly.

253°. *Theorem.*—Two triangles which have their vertices connecting concurrently have their corresponding sides intersecting collinearly. (Desargue's Theorem.)

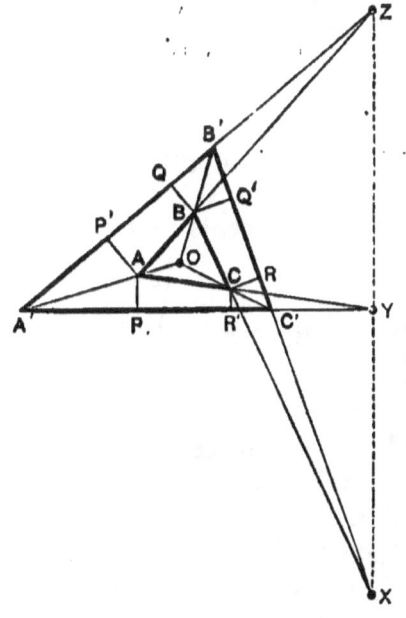

ABC, A'B'C' are two △s having their vertices connecting concurrently at O, and their corresponding sides intersecting in X, Y, Z. To prove that X, Y, Z are collinear.

Proof.—To the sides of △A'B'C' draw perpendiculars AP, AP', BQ, BQ', CR, CR'. Then, from similar △s,

$$\frac{BX}{CX} = \frac{BQ'}{CR'}$$

$$\frac{CY}{AY} = \frac{CR'}{AP'}$$

$$\frac{AZ}{BZ} = \frac{AP'}{BQ'}$$

∴ $\left(\frac{BX}{CX}\right) = \frac{AP'.BQ'.CR'}{AP.BQ.CR}$.

But $\frac{AP'}{AP} = \frac{\sin AA'B'}{\sin AA'C'}$

with similar expressions for the other ratios.

Also, since AA', BB', CC' are concurrent at O, they divide the angles A' B', C' so that

$$\frac{\sin AA'B' . \sin BB'C' . \sin CC'A'}{\sin AA'C' . \sin BB'A' . \sin CC'B'} = 1,$$

∴ $\left(\dfrac{BX}{CX}\right) = 1$, and X, Y, Z are collinear.

The converse of this theorem is readily proved, and will be left as an exercise to the reader.

Ex. A', B', C' are points upon the sides BC, CA, AB respectively of the △ABC, and AA', BB', CC' are concurrent in O. Then

1. AB and A'B', BC and B'C', CA and C'A' meet in three points Z, X, Y, which are collinear.

2. The lines AX, BY, CZ form a triangle with vertices A", B", C", such that AA", BB", CC" are concurrent in O.

OF RECTILINEAR FIGURES IN PERSPECTIVE.

254°. *Def.*—AB and A'B' are two segments and AA' and BB' meet in O.

Then the segments AB and A'B' are said to be in *perspective* at O, which is called their *centre of perspective.*

The term perspective is introduced from Optics, because an eye placed at O would see A' coinciding with A and B' with B, and the segment A'B' coinciding with AB.

By an extension of this idea O' is also a centre of perspective of AB and B'A'. O is then the *external* centre of perspective and O' is the *internal* centre.

Def.—Two rectilinear figures of the same number of sides are in *perspective* when every two corresponding sides have the same centre of perspective.

206 SYNTHETIC GEOMETRY.

Cor. 1. From the preceding definition it follows that two rectilinear figures of the same species are in perspective when the joins of their vertices, in pairs, are concurrent.

Cor. 2. When two triangles are in perspective, their vertices connect concurrently, and their corresponding sides intersect collinearly. (253°)

In triangles either of the above conditions is a criterion of the triangles being in perspective.

Def.—The line of collinearity of the intersections of corresponding sides of triangles in perspective is called their *axis of perspective;* and the point of concurrence of the joins of corresponding vertices is the *centre* of perspective.

255°. Let AA', BB', CC' be six points which connect concurrently in the order written.

These six points may be connected in four different ways so as to form pairs of triangles having the same centre of perspective, viz.,

ABC, A'B'C'; ABC', A'B'C; AB'C, A'BC'; A'BC, AB'C'.

These four pairs of conjugate triangles determine four axes of perspective, which intersect in six points; these points are centres of perspective of the sides of the two triangles taken in pairs, three X, Y, Z being external centres, and three X', Y', Z' being internal centres. (254°)

The points, the intersections which determine them, and the segments of which they are centres of perspective are given in the following table:—

Point.	Determined by Intersection of	Centre of Perspective to
X	BC – B'C'	BC' – B'C
Y	CA – C'A'	CA' – C'A
Z	AB – A'B'	AB' – A'B
X'	BC' – B'C	BC – B'C'
Y'	CA' – C'A	CA – C'A'
Z'	AB' – A'B	AB – A'B'

And the six points lie on the four lines thus,
$$XYZ,\ X'Y'Z,\ X'YZ',\ XY'Z'.$$

EXERCISES

1. The triangle formed by joining the centres of the three excircles of any triangle is in perspective with it.
2. The three chords of contact of the excircles of any triangle form a triangle in perspective with the original.
3. The tangents to the circumcircle of a triangle at the three vertices form a triangle in perspective with the original.

SECTION IV.

OF INVERSION AND INVERSE FIGURES.

256°. *Def.*—Two points so situated upon a centre-line of a circle that the radius is a geometric mean (169°, Def.) between their distances from the centre are called *inverse* points with respect to the circle.

Thus P and Q are inverse points if
$$CP \cdot CQ = CB^2 = R^2,$$
R being the radius.

The ⊙S is the *circle of inversion* or the inverting ⊙, and C is the *centre* of inversion.

Cor. From the definition :—

1. An indefinite number of pairs of inverse points may lie on the same centre-line.
2. An indefinite number of circles may have the same two points as inverse points.
3. Both points of a pair of inverse points lie upon the same side of the centre of inversion.
4. Of a pair of inverse points one lies within the circle and one without.

5. P and Q come together at B; so that any point on the circle of inversion is its own inverse.
6. When P comes to C, Q goes to ∞; so that the inverse of the centre of inversion is any point at infinity.

257°. *Problem.*—To find the circle to which two pairs of collinear points may be inverse points.

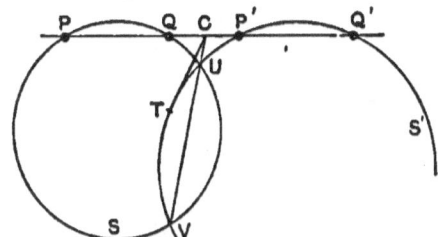

P, Q, P′, Q′ are the four collinear points, of which PQ and P′Q′ are respectively to be pairs of inverse points.

Through P, Q and through P′, Q′ describe any two circles S, S′ to intersect in two points U and V. Let the connector UV cut the axis of the points in C, and let CT be a tangent to circle S′. Then C is the centre and CT the radius of the required circle.

Proof.— $CT^2 = CP' \cdot CQ' = CU \cdot CV = CQ \cdot CP$.

Cor. If the points have the order P, P′, Q, Q′ the centre C is real and can be found as before, but it then lies within both circles S and S′, and no tangent can be drawn to either of these circles; in this case we say that the radius of the circle is imaginary although its centre is real. In the present case P and Q, as also P′ and Q′, lie upon opposite sides of C, and the rectangles CP.CQ and CP′.CQ′ are both negative. But R^2 being always positive (163°) cannot be equal to a negative magnitude.

When the points have the order P, P′, Q′, Q, the circle of inversion is again real.

Hence, in order that the circle of inversion may be real, each pair of points must lie wholly without the other, or one pair must lie between the others.

EXERCISES.

1. Given a ⊙ and a point without it to find the inverse point.
2. Given a ⊙ and a point within it to find the inverse point.
3. Given two points to find any ⊙ to which they shall be inverse.
4. In 3 the ⊙ is to have a given radius.
5. In 3 the ⊙ is to have a given centre on the line of the points.

258°. *Theorem.*—A ⊙ which passes through a pair of inverse points with respect to another ⊙ cuts the latter orthogonally. (115°, Defs. 1, 2)

And, conversely, a ⊙ which cuts another ⊙ orthogonally determines a pair of inverse points on any centre-line of the latter.

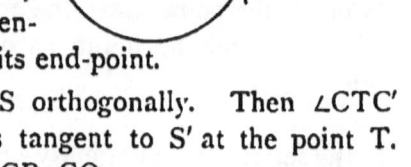

1. P and Q are inverse to ⊙S.
Then $CP \cdot CQ = CT^2$,
∴ CT is tangent to ⊙S'.
(176°, Cor. 3)
And ∴ S' cuts S orthogonally since the radius of S is perpendicular to the radius of S' at its end-point.

2. Conversely, let S' cut S orthogonally. Then ∠CTC' is a ⌐, and therefore CT is tangent to S' at the point T. Hence $CT^2 = CP \cdot CQ$,
and P and Q are inverse points to ⊙S.

Cor. 1. A ⊙ through a pair of points inverse to one another with respect to two ⊙s cuts both orthogonally.

Cor. 2. A ⊙ which cuts two ⊙s orthogonally determines on their common centre-line a pair of points which are inverse to one another with respect to both ⊙s.

Cor. 3. If the ⊙S cuts the ⊙s S' and S'' orthogonally, the tangents from the centre of S to the ⊙s S' and S'' are radii of S and therefore equal.

∴ (178°) a ⊙ which cuts two ⊙s orthogonally has its centre on their radical axis.

Cor. 4. A ⊙ having its centre on the radical axis of two given ⊙s, and cutting one of them orthogonally, cuts the other orthogonally also.

259°. Let P, Q be inverse points to circle S and D any point on it.

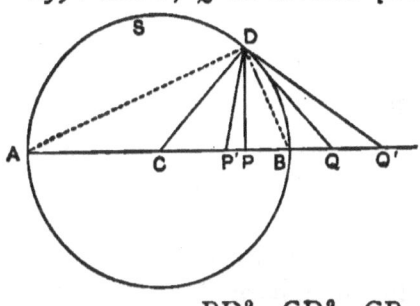

Then
$$CP \cdot CQ = CD^2,$$
∴ $CP : CD = CD : CQ$.
Hence the triangles CPD and CDQ are similar, and PD and DQ are homologous sides.

∴ $$\frac{PD^2}{QD^2} = \frac{CD^2}{CQ^2} = \frac{CP \cdot CQ}{CQ^2} = \frac{CP}{CQ}.$$

∴ the squares on the joins of any point on a circle with a pair of inverse points with respect to the circle are proportional to the distances of the inverse points from the centre.

Cor. 1. If P and Q are fixed, $PD^2 : QD^2$ is a fixed ratio.
∴ the locus of a point, for which the squares on its joins to two fixed points have a constant ratio, is a circle having the two fixed points as inverse points.

Cor. 2. When D comes to A and B we obtain
$$\frac{CP}{CQ} = \frac{PD^2}{QD^2} = \frac{PA^2}{QA^2} = \frac{PB^2}{QB^2},$$
∴ $$\frac{PD}{QD} = \frac{PA}{QA} = \frac{PB}{QB}.$$

Hence DA and DB are the bisectors of the ∠PDQ, and the segments PB and BQ subtend equal angles at D.

Hence the locus of a point at which two adjacent segments of the same line subtend equal angles is a circle passing through the common end-points of the segments and having their other end-points as inverse points.

Cor. 3. Let P', Q' be a second pair of inverse points. Then
$\angle BDP' = \angle BDQ'$, and $\angle BDP = \angle BDQ$,
∴ $\angle PDP' = \angle QDQ'$;
or the segments PP' and QQ' subtend equal angles at D.

Hence the locus of a point at which two non-adjacent segments of the same line subtend equal angles is a circle having the end-points as pairs of inverse points.

Cor. 4. Since $AP : AQ = PB : BQ$, (Cor. 1)
∴ P and Q divide the diameter AB in the same manner internally and externally, and B and A divide the segment PQ in the same manner internally and externally.

∴ from 208°, Cor., P, Q divide AB harmonically, and A, B divide PQ harmonically.

Hence, when two segments of the same line are such that the end-points of one divide the other harmonically, the circle on either segment as diameter has the end-points of the other segment as inverse points.

Exercises.

1. If a variable circle passes through a pair of inverse points with respect to a fixed circle, the common chord of the circles passes through a fixed point.
2. To draw a circle so as to pass through a given point and cut a given circle orthogonally.
3. To draw a circle to cut two given circles orthogonally.
4. On the common centre-line of two circles to find a pair of points which are inverse to both circles.

    Let C, C' be the centres of the circles S and S'. Take any point P, without both circles, and find its inverses P' and P" with respect to both circles. (257°, Ex. 1)

    The circle through P, P', and P" cuts the common centre-line CC' in the required points Q and Q'.
5. To describe a circle to pass through a given point and cut two given circles orthogonally.

6. To determine, on a given line, a point the ratio of whose distances from two fixed points is given.
7. To find a point upon a given line from which the parts of a given divided segment may subtend equal angles.
8. A, B, C, D are the vertices of a quadrangle. On the diagonal BD find a point at which the sides BA and BC subtend equal angles.
9. To draw a circle to pass through a given point and touch two given lines.

260°. *Def.*—One figure is the inverse of another when every point on one figure has its inverse upon the other figure.

Theorem.—The inverse of a circle is a circle when the centre of inversion is not on the figure to be inverted.

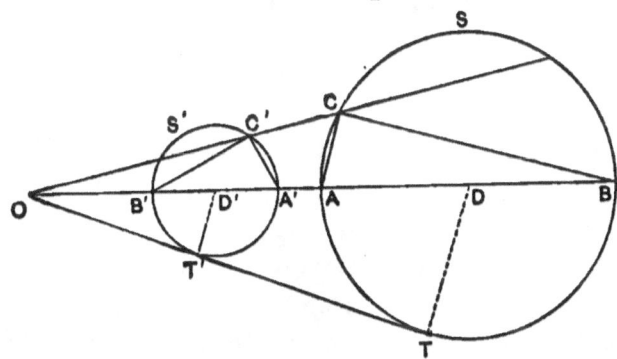

Let O be the centre of inversion and S be the circle to be inverted; and let A', B', C' be the inverses of A, B, C respectively. To prove that the ∠B'C'A'=⌐.

Proof.—$OA \cdot OA' = OB \cdot OB' = OC \cdot OC' = R^2$,   (256°)

∴   △OA'C' ≌ △OCA, and △OB'C' ≌ △OCB.

∴   (1) ∠OC'A'=∠OAC, and (2) ∠OC'B'=∠OBC.

And   ∠B'C'A'=∠OC'A'−OC'B'=∠OAC−∠OBC
              =∠ACB=⌐,

since ACB is in a semicircle.

∴ as C describes S, its inverse, C', describes the circle S' on A'B' as diameter.   *q.e.d.*

OF INVERSION AND INVERSE FIGURES. 213

Cor. 1. The point O is the intersection of common direct tangents.

Cor. 2. ∵ OD'. OD = OT . OT'. $\dfrac{OD'}{OT'} \cdot \dfrac{OD}{OT} = R^2 \cdot \dfrac{OD^2}{OT'^2}$,
where R is the radius of the circle of inversion;
∴ the centre of a circle and the centre of its inverse are not inverse points, unless OD = OT, *i.e.*, unless the centre of inversion is at ∞.

Cor. 3. When the circle to be inverted cuts the circle of inversion, its inverse cuts the circle of inversion in the same points. (256°, Cor. 5)

261°. *Theorem.*—A circle which passes through the centre of inversion inverts into a line.

Let O be the centre of inversion and S the circle to be inverted, and let P and P' be inverses of Q and Q'.

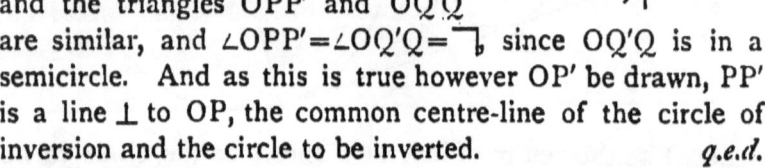

Proof.—Since
OP . OQ = OP' . OQ' = R²,
∴ OP : OP' = OQ' : OQ,
and the triangles OPP' and OQ'Q are similar, and ∠OPP' = ∠OQ'Q = ⌐, since OQ'Q is in a semicircle. And as this is true however OP' be drawn, PP' is a line ⊥ to OP, the common centre-line of the circle of inversion and the circle to be inverted. *q.e.d.*

Cor. 1. Since inversion is a reciprocal process, the inverse of a line is a circle through the centre of inversion and so situated that the line is ⊥ to the common centre-line of the two circles.

Cor. 2. Let I be the circle of inversion, and let PT and PT' be tangents to circles I and S respectively. Then,
PT² = OP² − OT² = OP² − OP . OQ = OP . PQ = PT'²,
∴ PT = PT',
∴ when a circle inverts into a line with respect to another circle, the line is the radical axis of the two circles. (178°, Def.)

Cor. 3. If a circle passes through the centre of inversion and cuts the circle of inversion, its inverse is their common chord.

Cor. 4. A centre-line is its own inverse.

Cor. 5. Considering the centre of inversion as a point-circle, its inverse is the line at ∞.

262°. A circle which cuts the circle of inversion orthogonally inverts into itself.

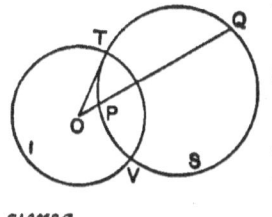

Since circle S cuts circle I orthogonally OT is a tangent to S, and hence
$$OP \cdot OQ = OT^2,$$
∴ P inverts into Q and Q into P, and the arc TQV inverts into TPV and *vice versa*.   *q.e.d.*

Cor. Since I cuts S orthogonally, it is evident that I inverts into itself with respect to S.

263°. A circle, its inverse, and the circle of inversion have a common radical axis.

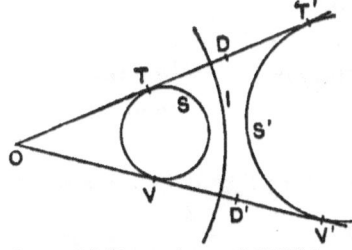

Let I be the circle of inversion, and let the circle S' be the inverse of S.

The tangents TT' and VV' meet at O (260°, Cor. 1), and T, T' are inverse points. D, the middle point of TT' is on the radical axis of S and S', and the circle with centre at D and radius DT cuts S and S' orthogonally. But this circle also cuts circle I orthogonally (258°). ∴ D is on the radical axes of I, S and S'.

Similarly D', the middle point of the tangent VV', is on the radical axes of I, S and S'.

∴ the three circles I, S, and S' have a common radical axis passing through D and D'.   *q.e.d.*

OF INVERSION AND INVERSE FIGURES. 215

Remarks.—This is proved more simply by supposing one of the circles to cut the circle of inversion. Then its inverse must cut the circle of inversion in the same points, and the common chord is the common radical axis.

The extension to cases of non-intersection follows from the law of continuity.

264°. *Theorem.*—The angle of intersection of two lines or circles is not changed in magnitude by inversion.

Let O be the centre of inversion, and let P be the point of intersection of two circles S and S', and Q its inverse. Take R and T points near P, and let U and V be their inverses. Then

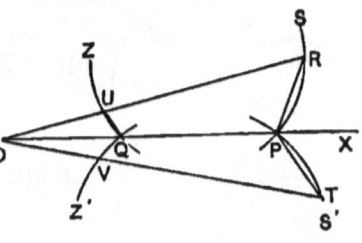

$$OU \cdot OR = OQ \cdot OP = OV \cdot OT = R^2,$$
∴   △OQU ≅ △ORP,
and   ∠OQU = ∠ORP = ∠RPX − ∠ROP.
Similarly   ∠OQV = ∠TPX − ∠TOP,
∴   ∠UQV = ∠RPT − ∠ROT.

But at the limit when R and T come to P the angle between the chords RP and PT becomes the angle between the circles (115°, Def. 1; 109°, Def. 1). And, since ∠ROT then vanishes, we have ultimately   ∠UQV = ∠RPT.

Therefore S and S', and their inverses Z and Z', intersect at the same angle.

Cor. 1. If two circles or a line and a circle touch one another their inverses also touch one another.

Cor. 2. If a circle inverts into a line, its centre-lines invert into circles having that line as a common diameter. For, since the circle cuts its centre-lines orthogonally, their inverses must cut orthogonally. But the centre-line is the only line cutting a circle orthogonally.

EXERCISES.

1. What is the result of inverting a triangle with respect to its incircle?
2. The circle of self-inversion of a given circle cuts it orthogonally.
3. Two circles intersect in P and Q, and AB is their common centre-line. What relation holds between the various parts when inverted with P as centre of inversion?
4. A circle cuts two circles orthogonally. Invert the system into two circles and their common centre-line.
5. Three circles cut each other orthogonally. If two be inverted into lines, their intersection is the centre of the third.

265°. The two following examples are important.

Ex. 1. Any two circles cut their common centre-line, and a circle which cuts them orthogonally in two sets of points which connect concurrently on the last-named circle.

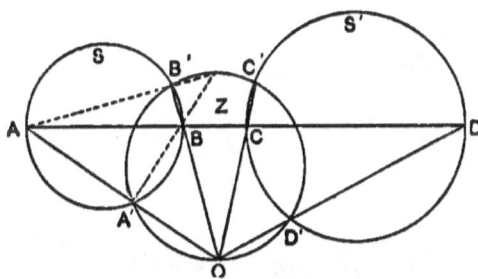

S and S' are the two given circles and Z a circle cutting them orthogonally.

Invert S and S' and their common centre-line with respect to a circle which cuts S and S' orthogonally and has its centre at some point O on Z. S and S' invert into themselves, and their centre-line into a circle through O cutting S and S' orthogonally, i.e., into circle Z.

∴ A' is the inverse of A, B' of B, etc., and the points AA', BB', CC', DD' connect concurrently at O.

Ex. 2. The nine-points circle of a triangle touches the incircle and the excircles of the triangle.

Let ABC be any triangle having its side AB touched by the incircle I at T, and by the ex-circle to the side c at T'. Take CH = CA and CD = CB, and join DH and HA.

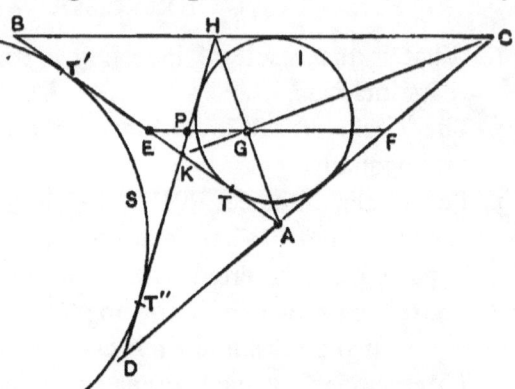

From the symmetry of the figure it is evident that HD touches both the circles I and S. Let E and F be the middle points of AB and AC, and let EF cut HA in G.

From 135°, Ex. 1, $\quad AT = BT' = s - a$,
whence $\quad ET = ET' = \tfrac{1}{2}(a-b)$.
But, since EF bisects HA, $EG = \tfrac{1}{2}BH = \tfrac{1}{2}(a-b)$,
∴ $\quad ET = ET' = EG$,

and the circle with E as centre and EG as radius cuts I and S orthogonally, and, with respect to this circle, the circles I and S invert into themselves.

Now, $\quad PF : HC = DF : DC = BC - CF : BC$,

∴ $$PF = \frac{HC}{BC}(BC - CF) = b - \frac{b^2}{2a},$$

∴ $$EP = EF - PF = \frac{a}{2} - b + \frac{b^2}{2a},$$

∴ $$EP \cdot EF = \left(\frac{a}{2} - b + \frac{b^2}{2a}\right)\frac{a}{2} = \tfrac{1}{4}(a-b)^2 = EG^2.$$

∴ P inverts into F, and the line HD into the circle through E and F, and by symmetry, through the middle point of BC. But this is the nine-points circle (116°, Ex. 6). And since HD touches I and S, the nine-points circle, which is the inverse of HD, touches the inverses of I and S, *i.e.*, I and S themselves.

And, similarly, the nine-points circle touches the two remaining excircles.

SECTION V.

OF POLE AND POLAR.

266°. *Def.*—The line through one of a pair of inverse points perpendicular to their axis is the *polar* of the other point with respect to the circle of inversion, and the point is the *pole* of the line.

The circle is called, in this relation, the *polar circle*, and its centre is called the *polar centre*.

From this definition and from the nature of inverse points we readily obtain the following :—

Cor. 1. The polar of the polar centre is a line at infinity. But, since the point which is the inverse of the centre may go to ∞ along any centre-line, all the lines obtained therefrom are polars of the centre. And as a point has in general but *one* polar with respect to any *one* circle, we speak of the polar of the centre as being *the* line at infinity, thus assuming that there is but *one* line at infinity.

Cor. 2. The polar of any point on the circle is the tangent at that point; or, a tangent to the polar circle is the polar to the point of contact.

Cor. 3. The pole of any line lies on that centre-line of the polar circle which is perpendicular to the former line.

Cor. 4. The pole of a centre-line of the polar circle lies at ∞ on the centre-line which is perpendicular to the former.

Cor. 5. The angle between the polars of two points is equal to the angle subtended by these points at the polar centre.

267°. *Theorem.*—If P and Q be two points, and P lies on the polar of Q, then Q lies on the polar of P.

OP and OQ are centre-lines of the polar circle I, and PE, ⊥ to OQ, is the polar of Q. To prove that QD, ⊥ to OP, is the polar of P.

Proof.—The △s ODQ and OEP are similar.

∴ OE : OP = OD : OQ,
and ∴ OE . OQ = OP . OD.
But E and Q are inverse points with respect to circle I,   (266°, Def.)
∴ P and D are inverse points,
and ∴ DQ is the polar of P.   *q.e.d.*

Def.—Points so related in position that each lies upon the polar of the other are *conjugate* points, and lines so related that each passes through the pole of the other are *conjugate* lines.

Thus P and Q are conjugate points and L and M are conjugate lines.

Cor. 1. If Q and, accordingly, its polar PV remain fixed while P moves along PE, L, which is the polar of P, will rotate about Q, becoming tangent to the circle when P comes to U or V, and cutting the circle when P passes without.

Similarly, if Q moves along L, M will rotate about the point P.

Cor. 2. As L will touch the circle at U and V, UV is the chord of contact for the point Q.

∴ for any point without a circle its chord of contact is its polar.

Cor. 3. For every position of P on the line M, its polar passes through Q.

∴ collinear points have their polars concurrent, and concurrent lines have their poles collinear, the point of concurrence being the pole of the line of collinearity.

EXERCISES.

1. Given a point and a line to find a circle to which they are pole and polar.
2. In Ex. 1 the circle is to pass through a given point.
3. In the figure of 267° trace the changes,
    (*a*) when P goes to ∞ along M ;
    (*b*) when P goes to ∞ along OD ;
    (*c*) when P moves along UV, what is the locus of D ?
4. From any point on a circle any number of chords are drawn, show that their poles all lie on the tangent at the point.
5. On a tangent to a circle any number of points are taken, show that all their polars with respect to the circle pass through the point of contact.

268°. *Theorem.*—The point of intersection of the polars of two points is the pole of the join of the points.

Let the polars of B and of C pass through A. Then A lies on the polar of B, and therefore B lies on the polar of A (267°). For similar reasons C lies on the polar of A.

∴ the polar of A passes through B and C and is their join.

q.e.d.

Cor. Let two polygons ABCD... and *abc*... be so situated that *a* is the pole of AB, *b* of BC, *c* of CD, etc.

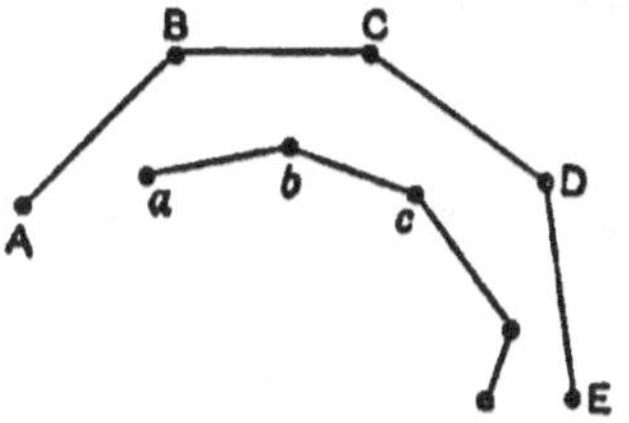

Then, since the polars of *a* and *b* meet at B, B is the pole of *ab* ; similarly C is the pole of *bc*, etc.

∴ if two polygons are such that the vertices of one are poles of the sides of the other, then, reciprocally, the vertices of the second polygon are poles of the sides of the first, the polar circle being the same in each case.

OF POLE AND POLAR.

Def. 1.--Polygons related as in the preceding corollary are *polar reciprocals* to one another.

Def. 2.—When two polar reciprocal △s become coincident, the resulting △ is *self-reciprocal* or *self-conjugate*, each vertex being the pole of the opposite side.

Def. 3.—The centre of the ⊙ with respect to which a △ is self-reciprocal is the polar centre of the △, and the ⊙ itself is the polar circle of the △.

269°. The orthocentre of a triangle is its polar centre.

Let ABC be a self-conjugate △. Then A is the pole of BC, and B of AC, and C of AB.

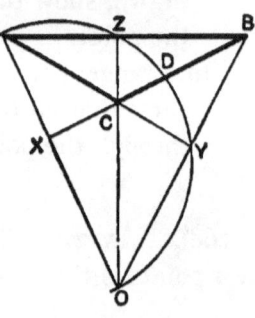

Let AX, ⊥ to BC, and BY, ⊥ to AC, meet in O. Then O is the orthocentre. (88°, Def.)

Now, as AX is ⊥ to BC, and as A is the pole of BC, the polar centre lies on AX. For similar reasons it lies on BY.   (266°, Cor. 3)

∴ O is the polar centre of the △ABC.   q.e.d.

Cor. 1. With respect to the polar ⊙ of the △, the ⊙ on AO as diameter inverts into a line ⊥ to AO (261°). And as A and X are inverse points, this line passes through X; therefore BC is the inverse of the ⊙ on OA as diameter.

Similarly, AC is the inverse of the ⊙ on OB as diameter, and AB of the ⊙ on OC as diameter.

Cor. 2. As the ⊙ on AO inverts into BC, the point D is inverse to itself, and is on the polar ⊙ of the △.   (256°, 5)

∴ OD is the polar radius of the △.

Cor. 3. If O falls within the △, it is evident that the ⊙ on OA as diameter will not cut CB. In this case the polar centre is real while the polar radius is imaginary. (257°, Cor.)

Hence a △ which has a real polar circle must be obtuse-angled.

Cor. 4. The ⊙ on BC as diameter passes through Y since Y is a ⌐.

But B and Y are inverse points to the polar ⊙.

∴ the polar ⊙ cuts orthogonally the ⊙ on BC as diameter. (258°)

Similarly for the circles on CA and AB.

∴ the polar ⊙ of a △ cuts orthogonally the circles having the three sides as diameters.

Cor. 5. The $\angle AOZ = \angle B$, $\angle BOZ = \angle A$, and $\angle OAC = \angle\left(C - \frac{\pi}{2}\right)$.

And $CX = OC \sin AOZ = OC \sin B$, also $= -b \cos C$,

∴ $OC = -\dfrac{b}{\sin B} \cdot \cos C = -d \cos C$,

where d is the diameter of the circumcircle (228°) to the triangles AOC or BOC or AOB or ABC, these being all equal. (116°, Ex. 4)

Similarly $OA = d \cos A$, $OB = d \cos B$.

But $OX = OC \cos B = -d \cos B \cos C$,

∴ $R^2 = OX \cdot OA = -d^2 \cos A \cos B \cos C$.

In order that the right-hand member may be +, one of the angles must be obtuse.

Cor. 6. $R^2 = OC \cdot OZ = OC(OC + CZ) = OC^2 + OC \cdot CZ$,

and $OC = -d \cos C$, and $CZ = a \sin B = \dfrac{ab}{d}$, (228°)

∴ $R^2 = d^2(1 - \sin^2 C) - ab \cos C$

$= d^2 - \tfrac{1}{2}(a^2 + b^2 + c^2)$. (217°)

If O is within the triangle, $d^2 < \tfrac{1}{2}(a^2 + b^2 + c^2)$ and R is imaginary.

EXERCISES.

1. If two triangles be polar reciprocals, the inverse of a side of one passes through a vertex of the other.
2. A right-angled triangle has its right-angled vertex at the centre of a polar circle. What is its polar reciprocal?
3. In Fig. of 269°, if the polar circle cuts CY produced in C′, prove that CY = YC′.

4. If P be any point, ABC a triangle, and A'B'C' its polar reciprocal with respect to a polar centre O, the perpendiculars from O on the joins PA, PB, and PC intersect the sides of A'B'C' collinearly.

270°. *Theorem.*—If two circles intersect orthogonally, the end-points of any diameter of either are conjugate points with respect to the other.

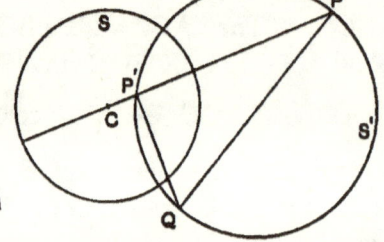

Let the circles S and S' intersect orthogonally, and let PQ be a diameter of circle S'. Then P' is inverse to P, and P'Q is ⊥ to CP.

∴ P'Q is the polar of P with respect to circle S.

∴ Q lies on the polar of P, and hence P lies on the polar of Q, and P and Q are conjugate points (267° and Def.).

q.e.d.

Cor. 1. $\quad PQ^2 = CP^2 + CQ^2 - 2CP \cdot CP' \quad\quad (172°, 2)$
$\quad\quad\quad\quad = CP^2 + CQ^2 - 2R^2$
$\quad\quad\quad\quad = CP^2 - R^2 + CQ^2 - R^2 = T^2 + T'^2,$

where T and T' are tangents from P and Q to the circle S.

∴ the square on the join of two conjugate points is equal to the sum of the squares on the tangents from these points to the polar circle.

Cor. 2. If a circle be orthogonal to any number of other circles, the end-points of any diameter of the first are conjugate points with respect to all the others. And when two points are conjugate to a number of circles the polars of either point with respect to all the circles pass through the other point.

271°. *Theorem.*—The distances of any two points from a polar centre are proportional to the distances of each point from the polar of the other with respect to that centre.

(Salmon.)

NN′ is the polar of P and MM′ is the polar of Q. Then
$$PO : QO = PM : QN.$$

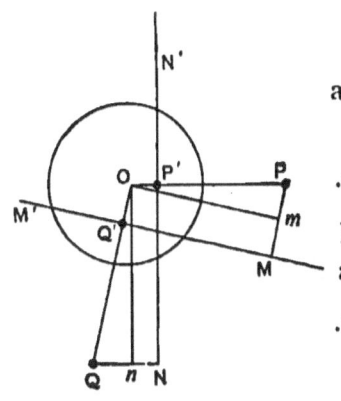

Proof.—Let O*m* be ∥ to MM′ and O*n* be ∥ to NN′. Then
$$OP'. OP = OQ'. OQ$$
$$\therefore \frac{OP}{OQ} = \frac{OQ'}{OP'} = \frac{Mm}{Nn}.$$

But the triangles OP*m* and OQ*n* are similar,
$$\therefore \frac{OP}{OQ} = \frac{Pm}{Qn} = \frac{Mm}{Nn}$$
$$= \frac{Pm + Mm}{Qn + Nn} = \frac{PM}{QN}. \quad q.e.d.$$

Cor. 1. A, B are any two points and L and M their polars, and P the point of contact of any tangent N.

AX and BY are ⊥ upon N, and PH and PK are ⊥ upon L and M respectively. Then
$$\frac{BY}{PK} = \frac{BO}{R} \text{ and } \frac{AX}{PH} = \frac{AO}{R},$$
$$\therefore \frac{BY \cdot AX}{PK \cdot PH} = \frac{AO \cdot BO}{R^2} = k, \text{ a constant,}$$
$$\therefore AX \cdot BY = k \cdot PH \cdot PK.$$

If A and B are on the circle, L and M become tangents having A and B as points of contact, and AO = BO = R.
$$\therefore AX \cdot BY = PH \cdot PK. \qquad \text{(See 211°, Ex. 1)}$$

Exercises.

1. If P and Q be the end-points of any diameter of the polar circle of the △ABC, the chords of contact of the point P with respect to the circles on AB, BC, and CA as diameters all pass through Q.

2. Two polar reciprocal triangles have their corresponding vertices joined. Of what points are these joins the polars?

OF POLE AND POLAR.

3. A, B, C are the vertices of a triangle and L, M, N the corresponding sides of its reciprocal polar. If T be a tangent at any point P, and AT is $\perp$ to T, etc.,
$$\frac{AT.BT.CT}{PL.PM.PN} = \frac{AO.BO.CO}{R^3} = \text{a constant.}$$
If A, B, C are on the circle,
$$AT.BT.CT = PL.PM.PN.$$

4. In Ex. 3, if A', B', C' be the vertices of the polar reciprocal,
$$\frac{A'T.B'T.C'T}{AT.BT.CT} = \frac{A'O.B'O.C'O}{R^3}.$$
The right-hand expression is independent of the position of T.

5. If ABC, A'B'C' be polar reciprocal triangles whose sides are respectively L, M, N and L', M', N', and if AM' is the $\perp$ from A to M', etc.,
$$AM'.BN'.CL' = AN'.BL'.CM',$$
and $$A'M.B'N.C'L = A'N.B'L.C'M.$$

272°. *Theorem.*—Triangles which are polar reciprocals to one another are in perspective.

Let ABC and A'B'C' be polar reciprocals. Let AP, AP' be perpendiculars on A'B' and A'C', BQ and BQ' be perpendiculars on B'C' and B'A', etc.

Then (271°) $\dfrac{AP'}{BQ} = \dfrac{AO}{BO}$, $\dfrac{BQ'}{CR} = \dfrac{BO}{CO}$, etc.,

$\therefore \quad \dfrac{AP'.BQ'.CR'}{AP.BQ.CR} = 1.$

But $AP' = AA' \sin AA'P$, $AP = AA' \sin AA'P'$,

$\therefore \quad \dfrac{AP'}{AP} = \dfrac{\sin AA'P'}{\sin AA'P'}$

and similarly for the other ratios. Hence AA', BB', CC' divide the angles at A, B, and C, so as to fulfil the criterion of 251°.

$\therefore$ AA', BB', and CC' are concurrent, and the triangles are in perspective. (254°, Cor. 2)

SECTION VI.

OF THE RADICAL AXIS.

273°. *Def.* 1.—The line perpendicular to the common centre-line of two circles, and dividing the distance between the centres into parts such that the difference of their squares is equal to the difference of the squares on the conterminous radii, is the radical axis of the two circles.

Cor. 1. When two circles intersect, their radical axis is the secant line through the points of intersection.

Cor. 2. When two circles touch, their radical axis is the common tangent at their point of contact.

Cor. 3. When two circles are mutually exclusive without contact, their radical axis lies between them.

Cor. 4. When two circles are equal and concentric, their radical axis is any line whatever, and when unequal and concentric it is the line at ∞.

Def. 2.—When three or more circles have a common radical axis they are said to be *co-axal*.

274°. If several circles pass through the same two points they form a co-axal system.

For (273°, Cor. 1) the line through the points is the radical axis of all the circles taken in pairs, and is therefore the common radical axis of the system.

Such circles are called circles of the *common point species*, contracted to *c.p.*-circles.

Let a system of *c.p.*-circles S, S_1, S_2, ..., pass through the common points P and Q, and let L'L be the right bisector of PQ.

Then the centres of all the circles of the *c.p.*-system lie on L'L and have M'M for their common radical axis.

Hence from any point C in M'M the tangents to all the circles are equal to one another. (178°)

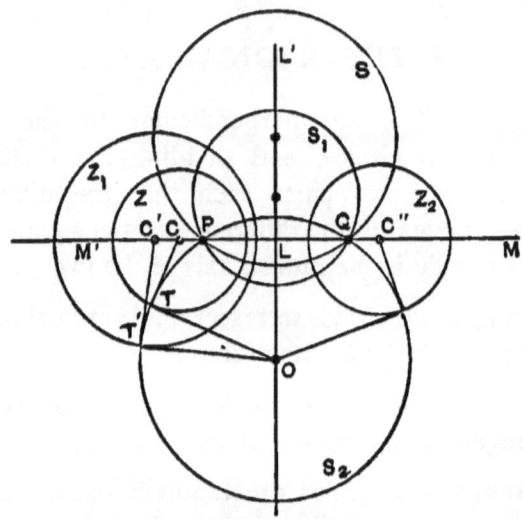

Let CT be one of these tangents. The circle Z with C as centre and CT as radius cuts all the *c.p.*-circles orthogonally.

Similarly, a system of circles Z, Z_1, Z_2, ... may be found with centres lying on M'M such that each one of the system cuts orthogonally every one of the *c.p.*-circles.

Since the centre of any circle of this new system is obtained by drawing a tangent from any one of the circles, as S_2, of the *c.p.*-species, to meet MM', it follows that no circle of this new system can have its centre lying between P and Q. As T approaches P the dependent circle Z contracts until it becomes the point-circle P, when T comes to coincidence with P.

Hence P and Q are limiting forms of the circles having their centres on M'M and cutting the *c.p.*-circles orthogonally. The circles of this second system are consequently called *limiting point circles*, contracted to *l.p.*-circles.

From the way in which *l.p.*-circles are obtained we see that from any point on L'L tangents to circles of the *l.p.*-system are all equal, and hence that L'L is the radical axis of the *l.p.*-circles. Thus the two systems of circles have their radical axes perpendicular, and every circle of one system cuts every circle of the other system orthogonally.

Hence P and Q are inverse points with respect to every circle of the *l.p.*-system, and with respect to any circle of either system all the circles of the other system invert into themselves.

If P and Q approach L, the *c.p.*-circles separate and the *l.p.*-circles approach, and when P and Q coincide at L the circles of both species pass through a common point, and the two radical axes become the common tangents to the respective systems.

If this change is continued in the same direction, P and Q become imaginary, and two new limiting points appear on the line L'L, so that the former *l.p.*-circles become *c.p.*-circles, and the former *c.p.*-circles become *l.p.*-circles.

Thus, in the systems under consideration, two limiting points are always real and two imaginary, except when they all become real by becoming coincident at L.

Cor. 1. As the *c.p.*-circles and the *l.p.*-circles cut each other orthogonally, the end-points of a diameter of any circle of one species are conjugate points with respect to every circle of the other species. But a circle of either species may be found to pass through any given point (259°, Ex. 5). ∴ the polars of a given point with respect to all the circles of either species are concurrent.

Cor. 2. Conversely, if the polars of a variable point P with respect to three circles are concurrent, the locus of the point is a circle which cuts them all orthogonally.

For let Q be the point of concurrence. Then P and Q are conjugate points with respect to each of the circles. Hence the circle on PQ as diameter cuts each of the circles orthogonally. (270°)

Cor. 3. If a system of circles is cut orthogonally by two circles, the system is co-axal.

For the centres of the cutting circles must be on the radical

OF THE RADICAL AXIS.

axis of all of the other circles taken in pairs; therefore they have a common radical axis.

Cor. 4. If two circles cut two other circles orthogonally, the common centre-line of either pair is the radical axis of the other pair.

Cor. 5. Two *l.p.*-circles being given, a circle of any required magnitude can be found co-axal with them. But if the circles be of the *c.p.*-species no circle can be co-axal with them whose diameter is less than the distance between the points.

EXERCISES.

1. Given two circles of the *l.p.*-species to find a circle with a given radius to be co-axal with them.
2. Given two circles of either species to find a circle to pass through a given point and be co-axal with them.
3. To find a point upon a given line or circle such that tangents from it to a given circle may be equal to its distance from a given point.
4. To find a point whose distances from two fixed points may be equal to tangents from it to two fixed circles.

275°. *Theorem.* —The difference of the squares on the tangents from any point to two circles is equal to twice the rectangle on the distance between the centres of the circles and the distance of the point from their radical axis.

Let P be the point, S and S' the circles, and LI their radical axis. Let PQ be $\perp$ to AB.

$PT^2 - PT'^2$

$\quad = PA^2 - PB^2 - (r^2 - r'^2)$,

where r, r' are radii of S and S'.

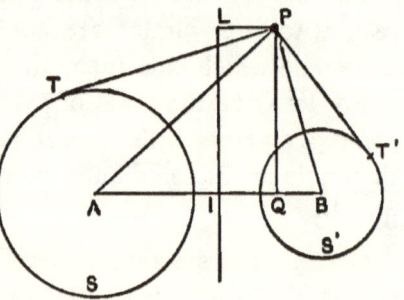

But, 273°, Def. 1,

$$r^2 - r'^2 = AI^2 - IB^2,$$

SYNTHETIC GEOMETRY.

and $\quad PA^2 - PB^2 = AQ^2 - QB^2,\qquad$ (172°, 1)

$\therefore\quad PT^2 - PT'^2 = AQ^2 - QB^2 - (AI^2 - IB^2)$
$= AB(AQ - QB) - AB(AI - IB)$
$= 2AB \cdot IQ = 2AB \cdot PL.\qquad$ q.e.d.

This relation is fundamental in the theory of the radical axis.

Cor. 1. When P is on the radical axis PL=0, and the tangents are equal, and when P is not on the radical axis the tangents are not equal.

Cor. 2. The radical axis bisects all common tangents to the two circles.

Cor. 3. If P lies on the circle S', PT'=0, and
$$PT^2 = 2AB \cdot PL,$$
$\therefore$ the square of the tangent from any point on one circle to another circle varies as the distance of the point from the radical axis of the circles.

Cor. 4. If C is the centre of a circle S'' passing through P and co-axal with S', $\quad PT^2 = 2AC \cdot PL.$

Now, if P could at any time leave this circle we would have
$$PT^2 - PT''^2 = 2AC \cdot PL,$$
where PT'' is the tangent from P to the circle S''
$\therefore\qquad PT^2 = PT^2 - PT''^2,$
which is impossible unless PT''=0.

Hence the locus of a point, which so moves that the square on the tangent from it to a given circle varies as the distance of the point from a given line, is a circle, and the line is the radical axis of this circle and the given circle.

Cor. 5. Let $PT' = k \cdot PT$, where k is a constant. Then
$$PT^2 - PT'^2 = (1 - k^2)PT^2 = 2AB \cdot PL,$$
$\therefore\qquad PT^2 = \dfrac{2AB \cdot PL}{1 - k^2}.$

As PT^2 varies as PL, P lies on a circle co-axal with S and S'.

$\therefore$ the locus of a point from which tangents to two given circles are in a constant ratio is a circle co-axal with the two.

EXERCISES.

1. In Cor. 5 what is the position of the locus for $k=0$, $k=1$, $k=>1$, k negative?
2. What is the locus of a point whose distances from two fixed points are in a constant ratio?
3. P and Q are inverse points to the circle I, and a line through P cuts circle I in A and B. PQ is the internal or external bisector of the $\angle AQB$, according as P is within or without the circle.
4. P, Q are the limiting points of the *l.p.*-circles S and S', and a tangent to S' at T cuts S in A and B.

    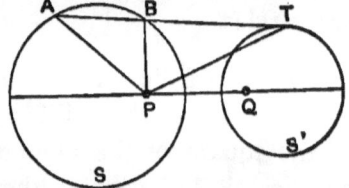

    Then, considering P as a point-circle, tangents from any point on S to P and S' are in a constant ratio.

    $\therefore$ AP : AT = BP : BT, and PT is the external bisector of $\angle$APB. If S' were enclosed by S, BT would be an internal bisector.
5. The points of contact of a common tangent to two *l.p.*-circles subtend a right angle at either limiting point.

276°. *Theorem*.—The radical axes of three circles taken in pairs are concurrent.

Let S_1, S_2, S_3 denote the circles, and let L be the radical axis of S_1 and S_2, M of S_2 and S_3, and N of S_3 and S_1.

L and M meet at some point O, from which $OT_1 = OT_2$, and $OT_2 = OT_3$, where OT_1 is the tangent from O to S_1, etc.,

$\therefore$ $OT_1 = OT_3$, and O is on N,

$\therefore$ L, M, and N are concurrent at O.

Def.—The point of concurrence of the three radical axes of three circles taken in pairs is called the *radical centre* of the circles.

Cor. 1. If S_1, S_2 are cut by a third circle Z, the common chords of S_1, Z and S_2, Z intersect on the radical axis of S_1 and S_2.

Hence to find the radical axis of two given circles S_1 and S_2, draw any two circles Z and Z_1 cutting the given circles. The chords S_1, Z and S_2, Z give one point on the radical axis and the chords S_1, Z_1 and S_2, Z_1 give a second point.

Cor. 2. If three circles intersect each other, their three common chords are concurrent. (See 249°, Ex. 1.)

Cor. 3. If a circle touches two others, the tangents at the points of contact meet upon the radical axis of the two.

Cor. 4. If a circle cuts three circles orthogonally, its centre is at their radical centre and its radius is the tangent from the radical centre to any one of them.

Cor. 5. If in Cor. 4 the three circles are co-axal, any number of circles may be found to cut them orthogonally, and hence they have no definite radical centre, as any point upon the common radical axis of the three becomes a radical centre.

Cor. 6. If in Cor. 4 the three circles mutually intersect one another, the radical centre is within each circle (Cor. 2), and no tangent can be drawn from the radical centre to any one of the circles. In this case the circle which cuts them all orthogonally has a real centre but an imaginary radius.

277°. *Theorem.*—If any three lines be drawn from the vertices of a $\triangle$ to the opposite sides, the polar centre of the $\triangle$ is the radical centre of the circles having these lines as diameters.

ABC is a $\triangle$ and O its orthocentre, and AP, BQ, CR are lines from the vertices to the opposite sides.

$\therefore \angle BXA = \rceil$,

the circle on AP as diameter passes through X, and OX . OA is equal to the square on the tangent from O to the circle on AP.

Similarly OY.OB is the square of the tangent from O to the circle on BQ as diameter, and similarly for OZ.OC. But as O is the polar centre of $\triangle ABC$, (269°)

∴ $\qquad$ OX.OA = OY.OB = OZ.OC.

∴ the tangents from O to the three circles on AP, BQ, and CR are equal, and O is their radical centre. $\qquad$ *q.e.d.*

Cor. 1. Let P, Q, R be collinear.

Then the polar centre of $\triangle ABC$ is the radical centre of circles on AP, BQ, and CR as diameters.

Again, in the $\triangle AQR$ AP, QB, and RC are lines from the vertices to the opposite sides.

∴ the polar centre of $\triangle AQR$ is the radical centre of circles on AP, BQ, and CR as diameters.

Similarly the polar centres of the $\triangle$s BPR and CPQ are radical centres to the same three circles.

But these $\triangle$s have not a common polar centre, as is readily seen. Hence the same three circles have four different radical centres. And this is possible only when the circles are co-axal. (276°, Cor. 5)

∴ the circles on AP, BQ, and CR are co-axal.

∴ if any three collinear points upon the sides of a $\triangle$ be joined with the opposite vertices, the circles on these joins as diameters are co-axal.

Cor. 2. Since ARPC is a quadrangle or tetragram (247°, Def. 2), and AP, BQ, CR are its three diagonals,

∴ the circles on the three diagonals of any quadrangle are co-axal.

Cor. 3. The middle points of AP, BQ, and CR are collinear. But ARPC is a quadrangle of which AP and CR are internal diagonals, and BQ the external diagonal.

∴ the middle points of the diagonals of a complete quadrangle, or tetragram, are collinear. (See 248°, Ex. 2)

Cor. 4. The four polar centres of the four triangles determined by the sides of a tetragram taken in threes are collinear

234   SYNTHETIC GEOMETRY.

and lie upon the common radical axis of the three circles having the diagonals of the tetragram as diameters.

278°. *Theorem.*—In general a system of co-axal circles inverts into a co-axal system of the same species.

(1.) Let the circles be of the *c.p.*-species.

The common points become two points by inversion, and the inverses of all the circles pass through them. Therefore the inverted system is one of *c.p.*-circles.

Cor. 1. The axis of the system (LL' of Fig. to 274°) inverts into a circle through the centre of inversion (261°, Cor. 1), and as all the inverted circles cut this orthogonally, the axis of the system and the two common points invert into a circle through the centre and a pair of inverse points to it.

(258°, Conv.)

Cor. 2. If one of the common points be taken as the centre of inversion, its inverse is at ∞.

The axis of the system then inverts into a circle through the centre of inversion, and having the inverse of the other common point as its centre, and all the circles of the system invert into centre-lines to this circle.

(2.) Let the circles be of the *l.p.*-species.

Let the circles S and S' pass through the limiting points and be thus *c.p.*-circles.

Generally S and S' invert into circles which cut the inverses of all the other circles orthogonally.   (264°)

∴ the intersections of the inverses of S and S' are limiting points, and the inverted system is of the *l.p.*-species.

Cor. 3. The axis of the system (MM' of Fig. to 274°) becomes a circle through the centre and passing through the limiting points of the inverted system, thus becoming one of the *c.p.*-circles of the system.

Cor. 4. If one of the limiting points be made the centre of

inversion, the circles S and S' become centre-lines, and the *l.p.*-circles become concentric circles.

Hence concentric circles are co-axal, their radical axis being at ∞.

EXERCISES.

1. What does the radical axis of (1, 278°) become?
2. What does the radical axis of (2, 278°) become?
3. How would you invert a system of concentric circles into a common system of *l.p.*-circles?
4. How would you invert a pencil of rays into a system of *c.p.*-circles.
5. The circles of 277° are common point circles.

279°. *Theorem.*—Any two circles can be inverted into equal circles.

Let S, S' be the circles having radii r and r', and let C, C' be the equal circles into which S and S' are to be inverted; and let the common radius be ρ.

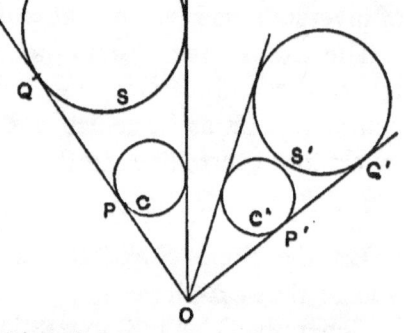

Then $\dfrac{OP}{OQ} = \dfrac{\rho}{r} = \dfrac{OP \cdot OQ}{OQ^2}$.

Similarly, $\dfrac{\rho}{r'} = \dfrac{OP' \cdot OQ'}{OQ'^2}$.

But, since P and Q and also P' and Q' are inverse points,
$$OP \cdot OQ = OP' \cdot OQ',$$
$$\therefore \quad \dfrac{OQ^2}{OQ'^2} = \dfrac{r}{r'} = \text{a constant},$$

and (275°, Cor. 5) O lies on a circle co-axal with S and S'. And with any point on this circle as a centre of inversion S and S' invert into equal circles.

Cor. 1. Any three non-co-axal circles can be inverted into equal circles.

For, let the circles be S, S', S", and let Z denote the locus of O for which S and S' invert into equal circles, and Z' the locus of O for which S and S" invert into equal circles. Then Z and Z' are circles of which Z is co-axal with S and S', and Z' is co-axal with S and S". And, as S, S', and S" are not co-axal, Z and Z' intersect in two points, with either of which as centre of inversion the three given circles can be inverted into equal circles.

Cor. 2. If S, S', and S" be *l.p.*-circles, Z and Z' being co-axal with them cannot intersect, and no centre exists with which the three given circles can be inverted into equal circles.

But if S, S' and S" be *c.p.*-circles, Z and Z' intersect in the common points, and the given circles invert into centre-lines of the circle of inversion, and having each an infinite radius these circles may be considered as being equal. (278°, Cor. 2)

Cor. 3. In general a circle can readily be found to touch three equal circles. Hence by inverting a system of three circles into equal circles, drawing a circle to touch the three, and then re-inverting we obtain a circle which touches three given circles.

If the three circles are co-axal, no circle can be found to touch the three.

280°. Let the circles S and S', with centres A and B and radii r and r, be cut by the circle Z with centre at O and radius OP=R. Let NL be the radical axis of S and S'.

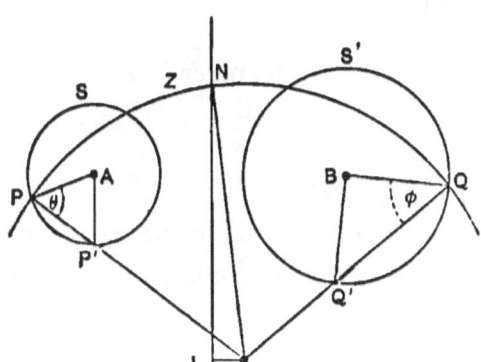

Since AP is ⊥ to the tangent at P to the circle S, and OP is ⊥ to the tangent at P to the circle Z, ∴ the ∠APO=θ is the angle of intersection of the circles S

OF THE RADICAL AXIS.

and Z ($115°$, Def. 1). Similarly BQO=ϕ is the angle of intersection of the circles S' and Z. Now

$$PP' = 2r\cos\theta = R - OP',$$
and $$QQ' = 2r'\cos\phi = R - OQ',$$
∴ $$OP' - OQ' = 2(r'\cos\phi - r\cos\theta).$$

But $R \cdot OP' - R \cdot OQ' = OT^2 - OT'^2$ (where OT is the tangent from O to S, etc.) $= 2AB \cdot OL$, \hfill ($275°$)

∴ $$R = \frac{AB}{r'\cos\phi - r\cos\theta} \cdot OL.$$

Cor. 1. When θ and ϕ are constant, R varies as OL.

∴ a variable circle which cuts two circles at constant angles has its radius varying as the distance of its centre from the radical axis of the circles.

Cor. 2. Under the conditions of Cor. 1 ON varies as OL, and ∴ $\dfrac{OL}{ON}$ is constant.

∴ a variable circle which cuts two circles at a constant angle cuts their radical axis at a constant angle.

Cor. 3. When OL=0, $r'\cos\phi = r\cos\theta$,
and $$r : r' = \cos\phi : \cos\theta.$$

∴ a circle with its centre on the radical axis of two other circles cuts them at angles whose cosines are inversely as the radii of the circles.

Cor. 4. If circle Z touches S and S', θ and ϕ are both zero or both equal to π, or one is zero and the other is π.

∴ when Z touches S and S', $R = \dfrac{AB}{\pm r' \mp r} \cdot OL$, where the variation in sign gives the four possible varieties of contact.

Cor. 5. When $\theta = \phi = \dfrac{\pi}{2}$, Z cuts S and S' orthogonally, and OL=0, and the centre of the cutting circle is on the radical axis of the two.

SECTION VII.

CENTRE AND AXES OF SIMILITUDE OR PERSPECTIVE.

The relations of two triangles in perspective have been given in Art. 254°. We here propose to extend these relations to the polygon and the circle.

281°. Let O, any point, be connected with the vertices A, B, C, ... of a polygon, and on OA, OB, OC, ... let points a, b, c, ... be taken so that

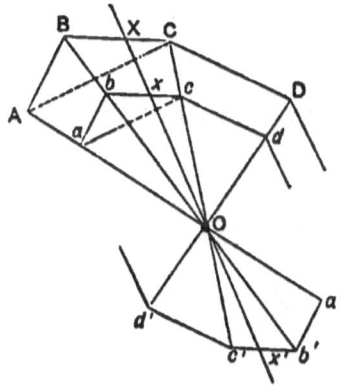

OA : Oa = OB : Ob = OC : Oc...
and
OA : Oa' = OB : Ob' = OC : Oc'...
Then, since OAB is a △ and ab is so drawn as to divide the sides proportionally in the same order,
∴ ab is ∥ to AB.   (202°, Conv.)
Similarly,
bc is ∥ to BC, cd to CD, etc.,
similarly, $b'c'$ is ∥ to BC, $c'd'$ to CD, etc.,
and   △OAB ≌ △Oab ≌ △O$a'b'$,
△OBC ≌ △Obc ≌ △O$b'c'$, ...
∴ the polygons ABC..., abc..., and $a'b'c'$... are all similar and have their homologous sides parallel.

Def.—The polygons ABCD... and $abcd$... are said to be *similarly* placed, and O is their *external* centre of similitude; while the polygons ABCD... and $a'b'c'd'$... are *oppositely* placed, and O is their *internal* centre of similitude.

Hence, when the lines joining any point to the vertices of a polygon are all divided in the same manner and in the same order, the points of division are the vertices of a second

CENTRE OF SIMILITUDE OR PERSPECTIVE. 239

polygon similar to the original, and so placed that the homologous sides of the two polygons are parallel.

282°. *When two similar polygons are so placed as to have their homologous sides parallel, they are in perspective, and the joins of corresponding vertices concur at a centre of similitude.*

Let ABCD..., *abcd*... be the polygons.

Since they are similar, AB : *ab* = BC : *bc* = CD : *cd*... (207°), and by hypothesis AB is ‖ to *ab*, BC to *bc*, etc.

Let A*a* and B*b* meet at some point O.

Then OAB is a △ and *ab* is ‖ to AB.

∴ $$\frac{OB}{Ob} = \frac{AB}{ab} = \frac{BC}{bc} = \text{etc.,}$$

∴ C*c* passes through O, and similarly D*d* passes through O, etc.

By writing $a'b'c'$... for *abc*... the theorem is proved for the polygon $a'b'c'd'$, which is oppositely placed to ABCD...

Cor. 1. If A*a* and B*b* meet at ∞, *ab* = AB, and hence *bc* = BC, etc., and the polygons are congruent.

Cor. 2. The joins of any two corresponding vertices as A, C ; *a*, *c* ; a', c' are evidently homologous lines in the polygons and are parallel.

Similarly any line through the centre O, as X*x*O*x'* is homologous for the polygons and divides them similarly.

283°. Let the polygon ABCD... have its sides indefinitely increased in number and diminished in length. Its limiting form (148°) is some curve upon which its vertices lie. A similar curve is the limiting form of the polygons *abcd*... as also of $a'b'c'd'$..., since every corresponding pair of limiting or vanishing elements are similar.

Hence, if two points on a variable radius vector have the ratio of their distances from the pole constant, the loci of the points are similar curves in perspective, and having the pole as a centre of perspective or similitude.

Cor. 1. In the limiting form of the polygons, the line BC becomes a tangent at B, and the line bc becomes a tangent at b. And similarly for the line $b'c'$.

∴ the tangents at homologous points on any two curves in perspective are parallel.

284°. Since $abcd$... and $a'b'c'd'$... are both in perspective with ABCD... and similar to it, we see that two similar polygons may be placed in two different relative positions so as to be in perspective, that is, they may be similarly placed or oppositely placed.

In a regular polygon of an even number of sides no distinction can be made between these two positions; or, two similar regular polygons are both similarly and oppositely placed at the same time when so placed as to be in perspective.

Hence two regular polygons of an even number of sides and of the same species, when so placed as to have their sides respectively parallel, have two centres of perspective, one due to the polygons being similarly placed, the external centre; and the other due to the polygons being oppositely placed, the internal centre.

Cor. Since the limiting form of a regular polygon is a circle (148°), two circles are always similarly and oppositely placed at the same time, and accordingly have always two centres of perspective or similitude.

285°. Let S and S' be two circles with centres C, C' and radii r, r' respectively, and let O and O' be their centres of perspective or similitude.

Let a secant line through O cut S in X and Y, and S' in X' and Y'.

CENTRE OF SIMILITUDE OR PERSPECTIVE. 241

Then O is the centre of similitude due to considering the

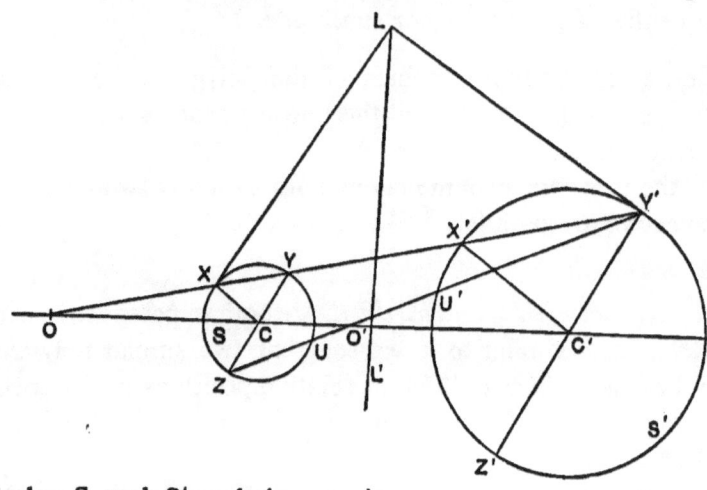

circles S and S' as being similarly placed.

Hence X and X', as also Y and Y', are homologous points, and (283°, Cor. 1) the tangents at X and X' are parallel. So also the tangents at Y and Y' are parallel.

Again O' is the centre of similitude due to considering the circles as being oppositely placed,

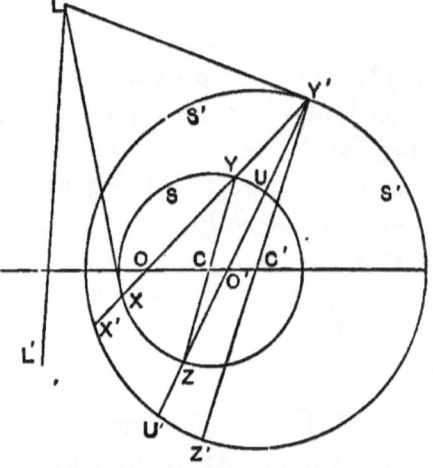

and for this centre Z and Y' as also U and U' are homologous points; and tangents at Y' and Z are parallel, and so also are tangents at U and U'.

Hence YZ is a diameter of the circle S and is parallel to Y'Z' a diameter of the circle S'.

Hence to find the centres of similitude of two given circles:—Draw parallel diameters, one to each circle, and connect their end-points directly and transversely. The direct connector cuts the common centre-line in the external

Q

centre of similitude, and the transverse connector cuts it in the internal centre of similitude.

286°. Since $OX : OX' = OY : OY'$, if X and Y become coincident, X' and Y' become coincident also.

∴ a line through O tangent to one of the circles is tangent to the other also, or O is the point where a common tangent cuts the common centre-line. A similar remark applies to O'.

When the circles exclude one another the centres of similitude are the intersections of common tangents of the same name, direct and transverse.

When one circle lies within the other (2nd Fig.) the common tangents are imaginary, although O and O' their points of intersection are real.

287°. Since $\triangle OCY \backsimeq \triangle OC'Y'$, ∴ $OC : OC' = r : r'$,
and since $\triangle O'CZ \backsimeq \triangle O'C'Y'$, ∴ $O'C : O'C' = r : r'$.

∴ the centres of similitude of two circles are the points which divide, externally and internally, the join of the centres of the circles into parts which are as the conterminous radii.

The preceding relations give

$$OC = \frac{r}{r'-r} \cdot CC', \text{ and } O'C = \frac{r}{r'+r} \cdot CC'.$$

∴ OC is $\gtreqless r$ according as CC' is $\gtreqless r'-r$,
and $O'C$ is $\gtreqless r$ according as CC' is $\gtreqless r'+r$.

Hence

1. O lies within the circle S when the distance between the centres is less than the difference of the radii, and O' lies within the circle S when the difference between the centres is less than the sum of the radii.

2. When the circles exclude each other without contact both centres of similitude lie without both circles.

3. When the circles touch externally, the point of contact is the internal centre of similitude.

4. When one circle touches the other internally, the point of contact is the external centre of similitude.

CENTRE OF SIMILITUDE OR PERSPECTIVE.

5. When the circles are concentric, the centres of similitude coincide with the common centre of the circles, unless the circles are also equal, when one centre of similitude becomes any point whatever.

6. If one of the circles becomes a point, both centres of similitude coincide with the point.

288°. Def.—The circle having the centres of similitude of two given circles as end-points of a diameter is called the *circle of similitude* of the given circles.

The contraction ⊙ *of s.* will be used for circle of similitude.

Cor. 1. Let S, S' be two circles and Z their ⊙ *of s.*

Since O and O' are two points from which tangents to circles S and S' are in the constant ratio of r to r', the circle Z is co-axal with S and S' (275°, Cor. 5). Hence any two circles and their ⊙ *of s.* are co-axal.

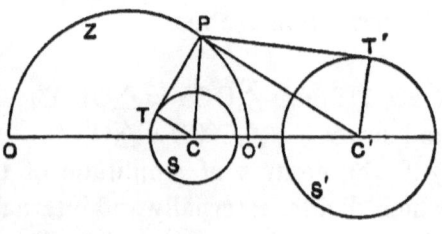

Cor. 2. From any point P on circle Z,
$$PT : TC = PT' : T'C',$$
and ∴ $\angle TPC = \angle T'PC'.$

Hence, at any point on the ⊙ *of s.* of two circles, the two circles subtend equal angles.

Cor. 3. $OC = CC' \cdot \dfrac{r}{r'-r}$, and $O'C = CC' \cdot \dfrac{r}{r'+r}$. (287°)

whence $OO' = CC' \cdot \dfrac{2rr'}{r'^2 - r^2}.$

∴ 1. The ⊙ *of s.* is a line, the radical axis, when the given circles are equal ($r = r'$).

2. The ⊙ *of s.* becomes a point when one of the two given circles becomes a point (r or $r' = 0$).

3. The ⊙ *of s.* is a point when the given circles are concentric ($CC' = 0$).

289°. Def.—With reference to the centre O (Fig. of 285°), X and Y′, as also X′ and Y, are called *antihomologous* points. Similarly with respect to the centre O′, U′ and Z, as also U and Y′, are antihomologous points.

Let tangents at X and Y′ meet at L. Then, since CX is ∥ to C′X′, ∠CXY=∠C′X′Y′=∠C′Y′X′. But ∠LXY is comp. of ∠CXY and ∠LY′X′ is comp. of ∠C′Y′X′.

∴ △LXY′ is isosceles, and LX=LY′.

∴ L is on the radical axis of S and S′.

Similarly it may be proved that pairs of tangents at Y and X′, at U and Y′, and at U′ and Z, meet on the radical axis of S and S′, and the tangent at U passes through L.

∴ tangents at a pair of antihomologous points meet on the radical axis.

Cor. 1. The join of the points of contact of two equal tangents to two circles passes through a centre of similitude of the two circles.

Cor. 2. When a circle cuts two circles orthogonally, the joins of the points of intersection taken in pairs of one from each circle pass through the centres of similitude of the two circles.

290°. Since $OX : OX' = r : r'$,

∴ $OX \cdot OY' : OX' \cdot OY' = r : r'$.

But $OX' \cdot OY' =$ the square of the tangent from O to the circle S′ and is therefore constant.

∴ $OX \cdot OY' = \dfrac{r}{r'} \cdot OT'^2 =$ a constant.

∴ X and Y′ are inverse points with respect to a circle whose centre is at O and whose radius is $OT'\sqrt{\dfrac{r}{r'}}$.

Def.—This circle is called the circle of *antisimilitude*, and will be contracted to ⊙ *of ans.*

Evidently the circles S and S′ are inverse to one another with respect to their ⊙ *of ans*.

For the centre O' the product $OU \cdot OY'$ is negative, and the ⊙ *of ans.* corresponding to this centre is imaginary.

Cor. 1. Denoting the distance CC' by d, and the difference between the radii $(r'-r)$ by δ, we have
$$R^2 = rr' \cdot \frac{d^2 - \delta^2}{\delta^2},$$
where R = the radius of the ⊙ *of ans.* Hence
1. When either circle becomes a point their ⊙ *of ans.* becomes a point.
2. When the circles S and S' are equal, the ⊙ *of ans.* becomes the radical axis of the two circles.
3. When one circle touches the other internally the ⊙ *of ans.* becomes a point-circle. ($d=\delta$.)
4. When one circle includes the other without contact the ⊙ *of ans.* is imaginary. ($d<\delta$.)

Cor. 2. Two circles and their circle of antisimilitude are co-axal. (263°)

Cor. 3. If two circles be inverted with respect to their circle of antisimilitude, they exchange places, and their radical axis being a line circle co-axal with the two circles becomes a circle through O co-axal with the two.

The only circle satisfying this condition is the circle of similitude of the two circles. Therefore the radical axis inverts into the circle of similitude, and the circle of similitude into the radical axis.

Hence every line through O cuts the radical axis and the circle of similitude of two circles at the same angle.

291°. *Def.*—When a circle touches two others so as to exclude both or to include both, it is said to *touch them similarly*, or to have *contacts of like kind* with the two. When it includes the one and excludes the other, it is said to *touch them dissimilarly*, or to have *contacts of unlike kinds* with the two.

246   SYNTHETIC GEOMETRY.

292°. *Theorem.*—When a circle touches two other circles, its chord of contact passes through their external centre of similitude when the contacts are of like kind, and through their internal centre of similitude when the contacts are of unlike kinds.

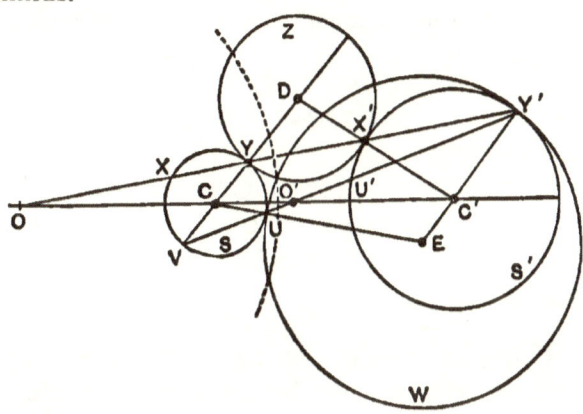

Proof.—Let circle Z touch circles S and S' at Y and X'. Then CYD and C'X'D are lines.   (113°, Cor. 1)

Let XYX'Y' be the secant through Y and X'. Then

$$\angle CXY = \angle CYX = \angle DYX' = \angle DX'Y = \angle CX'Y'.$$

∴ CX and C'X' are parallel, and X'X passes through the external centre of similitude O.   (285°)

Similarly, if Z' includes both S and S', it may be proved that its chord of contact passes through O.

Again, let the circle W, with centre E, touch S' at Y' and S at U so as to include S' and exclude S, and let UY' be the chord of contact. Then

$$\angle CVU = \angle CUV = \angle EUY' = \angle EY'U,$$

∴ EY' and CV are parallel and VY' connects them transversely; ∴ VY' passes through O'.   *q.e.d.*

Cor. 1. Every circle which touches S and S' similarly is cut orthogonally by the external circle of antisimilitude of S and S'.

Cor. 2. If two circles touch S and S' externally their points of contact are concyclic. (116°, Ex. 2)

But the points of contact of either circle with S and S' are antihomologous points to the centre O.

∴ if a circle cuts two others in a pair of antihomologous points it cuts them in a second pair of antihomologous points.

Cor. 3. If two circles touch two other circles similarly, the radical axis of either pair passes through a centre of similitude of the other pair.

For, if Z and Z' be two circles touching S and S' externally, the external circle of antisimilitude of S and S' cuts Z and Z' orthogonally (Cor. 1) and therefore has its centre on the radical axis of Z and Z'.

Cor. 4. If any number of circles touch S and S' similarly, they are all cut orthogonally by the external circle of antisimilitude of S and S', and all their chords of contact and all their chords of intersection with one another are concurrent at the external centre of antisimilitude of S and S'.

293°. *Theorem.*—If the circle Z touches the circles S and S', the chord of contact of Z and the radical axis of S and S' are conjugate lines with respect to the circle Z.

Proof.—Let Z touch S and S' in Y and X' respectively. The tangents at Y and X' meet at a point P on the radical axis of S and S'. (178°)

But P is the pole of the chord of contact YX'.

∴ the radical axis passes through the pole of the chord of contact, and reciprocally the chord of contact passes through the pole of the radical axis (267°, Def.) and the lines are conjugate. *q.e.d.*

AXES OF SIMILITUDE.

294°. Let S_1, S_2, S_3 denote three circles having their centres A, B, C and radii r_1, r_2, r_3, and let X, X', Y, Y', Z, Z' be their six centres of similitude.

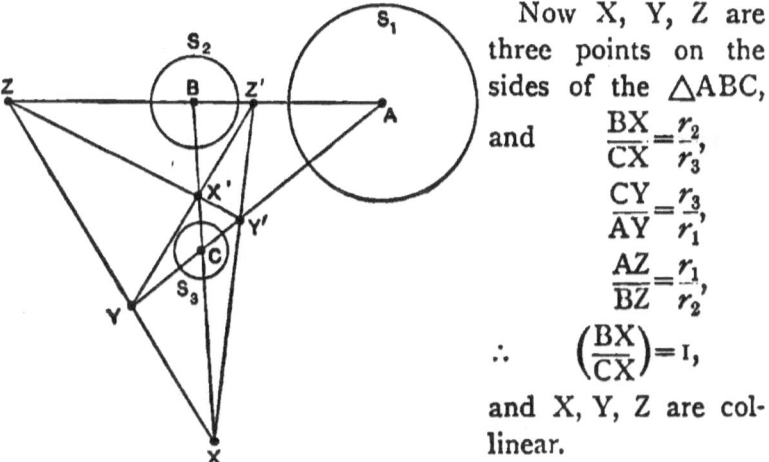

Now X, Y, Z are three points on the sides of the $\triangle ABC$, and
$$\frac{BX}{CX} = \frac{r_2}{r_3},$$
$$\frac{CY}{AY} = \frac{r_3}{r_1},$$
$$\frac{AZ}{BZ} = \frac{r_1}{r_2},$$
$$\therefore \quad \left(\frac{BX}{CX}\right) = 1,$$
and X, Y, Z are collinear.

Similarly it is proved that the triads of points XY'Z', YZ'X', ZX'Y' are collinear.

Def.—These lines of collinearity of the centres of similitude of the three circles taken in pairs are the *axes* of similitude of the circles. The line XYZ is the external axis, as being external to all the circles, and the other three, passing between the circles, are internal axes.

Cor. 1. If an axis of similitude touches any one of the circles it touches all three of them. (286°)

Cor. 2. If an axis of similitude cuts any one of the circles it cuts all three at the same angle, and the intercepted chords are proportional to the corresponding radii.

Cor. 3. Since XYX'Y' is a quadrangle whereof XX', YY', and ZZ' are the three diagonals, the circles on XX', YY', and ZZ' as diameters are co-axal. (277°, Cor. 2)

$\therefore$ the circles of similitude of three circles taken in pairs are co-axal.

AXES OF SIMILITUDE OR PERSPECTIVE. 249

Cor. 4. Since the three circles of similitude are of the *c.p.*-species, two points may be found from which any three circles subtend equal angles. These are the common points to the three circles of similitude. (288°, Cor. 2)

Cor. 5. The groups of circles on the following triads of segments as diameters are severally co-axal,
AX, BY, CZ; AX, YZ', Y'Z; BY, Z'X, ZX'; CZ, XY', X'Y.

295°. Any two circles Z and Z', which touch three circles S_1, S_2, S_3 similarly, cut their circles of antisimilitude orthogonally (292°, Cor. 1), and therefore have their centres at the radical centre of the three circles of antisimilitude.
(276°, Cor. 4)

But Z and Z' have not necessarily the same centre.
∴ the three circles of antisimilitude of the circles S_1, S_2, and S_3 are co-axal, and their common radical axis passes through the centres of Z and Z'.

296°. *Theorem.*—If two circles touch three circles similarly, the radical axis of the two is an axis of similitude of the three; and the radical centre of the three is a centre of similitude of the two.

Proof.—The circles S and S' touch the three circles A, B, and C similarly.

1. Since S and S' touch A and B similarly, the radical axis of S and S' passes through a centre of similitude of A and B.
(292°, Cor. 3)

Also, the radical axis of S and S' passes through a centre of similitude of B and C, and through a centre of similitude of C and A.

∴ the radical axis of S and S' is an axis of similitude of the three circles A, B, and C.

2. Again, since A and B touch S and S', the radical axis of A and B passes through a centre of similitude of S and S'.

For similar reasons, and because A, B, and C touch S and S' similarly, the radical axes of B and C, and of C and A, pass through the same centre of similitude of S and S'. But these three radical axes meet at the radical centre of A, B, and C.

∴ the radical centre of A, B, and C is a centre of similitude of S and S'. *q.e.d.*

297°. *Problem.*—To construct a circle which shall touch three given circles.

In the figure of 296°, let A, B, and C be the three given circles, and let S and S' be two circles which are solutions of the problem.

Let L denote one of the axes of similitude of A, B, and C, and let O be their radical centre. These are given when the circles A, B, and C are given.

Now L is the radical axis of S and S' (296°, 1), and O is one of their centres of similitude.

But as A touches S and S' the chord of contact of A passes through the pole of L with respect to A (293°). Similarly the chords of contact of B and C pass through the poles of L with respect to B and C respectively. And these chords are concurrent at O. (292°)

Hence the following construction :—

Find O the radical centre and L an axis of similitude of A, B, and C. Take the poles of L with respect to each of these circles, and let them be the points p, q, r respectively.

Then Op, Oq, Or are the chords of contact for the three given circles, and three points being thus found for each of two touching circles, S and S', these circles are determined.

(This elegant solution of a famous problem is due to M. Gergonne.)

Cor. As each axis of similitude gives different poles with respect to A, B, and C, while there is but one radical centre

AXES OF SIMILITUDE OR PERSPECTIVE. 251

O, in general each axis of similitude determines two touching circles; and as there are four axes of similitude there are eight circles, in pairs of twos, which touch three given circles.

Putting i and e for internal and external contact with the touching circle, we may classify the eight circles as follows:

(See 294°)

Axes of Similitude.	A	B	C	
X Y Z	i	i	i	} 1 pr.
	e	e	e	
X Y′ Z′	e	i	i	} 2 pr.
	i	e	e	
X′ Y Z′	i	e	i	} 3 pr.
	e	i	e	
X′ Y′ Z	i	i	e	} 4 pr.
	e	e	i	

PART V.

ON HARMONIC AND ANHARMONIC RATIOS—HOMOGRAPHY, INVOLUTION, ETC.

SECTION I.

GENERAL CONSIDERATIONS IN REGARD TO HARMONIC AND ANHARMONIC DIVISION.

298°. Let C be a point dividing a segment AB. The position of C in relation to A and B is determined by the ratio AC : BC. For, if we know this ratio, we know completely the position of C with respect to A and B. If this ratio is negative, C lies between A and B; if positive, C does not lie between A and B. If AC : BC = -1, C is the internal bisector of AB; and if AC : BC = $+1$, C is the external bisector of AB, *i.e.*, a point at ∞ in the direction AB or BA.

Let D be a second point dividing AB. The position of D is known when the ratio AD : BD is known.

Def.—If we denote the ratio AC : BC by *m*, and the ratio AD : BD by *n*, the two ratios *m* : *n* and *n* : *m*, which are reciprocals of one another, are called the two *anharmonic ratios* of the division of the segment AB by the points C and D, or the harmonoids of the range A, B, C, D.

HARMONIC AND ANHARMONIC DIVISION. 253

Either of the two anharmonic ratios expresses a relation between the parts into which the segment AB is divided by the points C and D.

Evidently the two anharmonic ratios have the same sign, and when one of them is zero the other is infinite, and *vice versa*.

These ratios may be written:—

1. $\dfrac{AC}{BC} : \dfrac{AD}{BD}$ or $\dfrac{AC \cdot BD}{BC \cdot AD}$ or $\dfrac{AC \cdot BD}{AD \cdot BC}$.

2. $\dfrac{AD}{BD} : \dfrac{AC}{BC}$ or $\dfrac{AD \cdot BC}{BD \cdot AC}$ or $\dfrac{AD \cdot BC}{AC \cdot BD}$.

The last form is to be preferred, other things being convenient, on account of its symmetry with respect to A and B, the end-points of the divided segment.

299°. The following results readily follow.

1. Let $\dfrac{AC \cdot BD}{AD \cdot BC}$ be +. Then $\dfrac{AC}{BC}$ and $\dfrac{AD}{BD}$ have like signs, and therefore C and D both divide AB internally or both externally. (298°)

In this case the order of the points must be some one of the following set, where AB is the segment divided, and the letters C and D are considered as being interchangeable:

CDAB, ACDB, CABD, ABCD.

2. Let $\dfrac{AC \cdot BD}{AD \cdot BC}$ be −. Then $\dfrac{AC}{BC}$ and $\dfrac{AD}{BD}$ have opposite signs, and one point divides AB internally and the other externally.

The order of the points is then one of the set CADB, ACBD.

3. When either of the two anharmonic ratios is ±1, these ratios are equal.

4. Let $\dfrac{AC \cdot BD}{AD \cdot BC} = +1$. Then $\dfrac{AC}{BC} = \dfrac{AD}{BD}$, and C and D are both internal or both external.

Also $\dfrac{AC-BC}{BC} = \dfrac{AD-BD}{BD}$, or $\dfrac{AB}{BC} = \dfrac{AB}{BD}$,
and C and D coincide.

Hence, when C and D are distinct points, the anharmonic ratio of the parts into which C and D divide AB cannot be positive unity.

5. Let $\dfrac{AC \cdot BD}{AD \cdot BC} = -1$. Then $\dfrac{AC}{BC} = -\dfrac{AD}{BD}$.

And since C and D are now one external and one internal (2), they divide the segment AB in the same ratio internally and externally, disregarding sign. Such division of a line segment is called *harmonic*. (208°, Cor. 1)

Harmonic division and harmonic ratio have been long employed, and from being only a special case of the more general ratio, this latter was named "anharmonic" by Chasles, "who was the first to perceive its utility and to apply it extensively in Geometry."

300°. *Def.*—When we consider AB and CD as being two segments of the same line we say that CD divides AB, and that AB divides CD.

Now the anharmonic ratios in which CD divides AB are
$$\dfrac{AC \cdot BD}{AD \cdot BC} \text{ and } \dfrac{AD \cdot BC}{AC \cdot BD}.$$
And the anharmonic ratios in which AB divides CD are
$$\dfrac{CA \cdot DB}{CB \cdot DA} \text{ and } \dfrac{CB \cdot DA}{CA \cdot DB}.$$
But the anharmonic ratios of these sets are equal each to each in both sign and magnitude.

∴ *the anharmonic ratios in which* CD *divides* AB *are the same as those in which* AB *divides* CD.

Or, *any two segments of a common line divide each other equianharmonically.*

301°. Four points A, B, C, D taken on a line determine six segments AB, AC, AD, BC, BD, and CD.

These may be arranged in three groups of two each, so that in each group one segment may be considered as dividing the others, viz., AB, CD ; BC, AD ; CA, BD.

Each group gives two anharmonic ratios, reciprocals of one another; and thus the anharmonic ratios determined by a range of four points, taken in all their possible relations, are six in number, of which three are reciprocals of the other three.

These six ratios are not independent, for, besides the reciprocal relations mentioned, they are connected by three relations which enable us to find all of them when any one is given.

Denote $\dfrac{AC \cdot BD}{AD \cdot BC}$ by P, $\dfrac{BA \cdot CD}{BD \cdot CA}$ by Q, $\dfrac{CB \cdot AD}{CD \cdot AB}$ by R.

Then P, Q, R are the anharmonic ratios of the groups ABCD, BCAD, and CABD, each taken in the same order.

But in any range of four (233°) we have

$$AB \cdot CD + BC \cdot AD + CA \cdot BD = 0.$$

And dividing this expression by each of its terms in succession, we obtain $\quad Q + \dfrac{1}{P} = R + \dfrac{1}{Q} = P + \dfrac{1}{R} = 1.$

From the symmetry of these relations we infer that any general properties belonging to one couple of anharmonic ratios, consisting of any ratio and its reciprocal, belong equally to all.

Hence the properties of only one ratio need be studied.

The symbolic expression {ABCD} denotes any one of the anharmonic ratios, and may be made to give all of them by reading the constituent letters in all possible orders.

Except in the case of harmonic ratio, or in other special cases, we shall read the symbol in the one order of alternating the letters in the numerator and grouping the extremes and means in the denominator. Thus

$$\{ABCD\} \text{ denotes } \dfrac{AC \cdot BD}{AD \cdot BC}.$$

It is scarcely necessary to say that whatever order may be

adopted in reading the symbol, the *same* order must be employed for *each* when comparing two symbols.

302°. *Theorem.*—Any two constituents of the anharmonic symbol may be interchanged if the remaining two are interchanged also, without affecting the value of the symbol.

Proof.— $\{ABCD\} = AC.BD : AD.BC$.
Interchange any two as A and C, and also interchange the remaining two B and D. Then
$$\{CDAB\} = CA.DB : CB.DA$$
$$= AC.BD : AD.BC.$$
Similarly it is proved that
$$\{ABCD\} = \{BADC\} = \{CDAB\} = \{DCBA\}. \quad q.e.d.$$

303°. If interchanging the first two letters, or the last two, without interchanging the remaining letters, does not alter the value of the ratio, it is harmonic.

For, let $\{ABCD\} = \{ABDC\}.$
Then $$\frac{AC.BD}{AD.BC} = \frac{AD.BC}{AC.BD},$$
or, multiplying across and taking square roots,
$$AC.BD = \pm AD.BC.$$
But the positive value must be rejected (299°, 4), and the negative value gives the condition of harmonic division.

304°. Let ABCD be any range of four and O any point not on its axis.

The anharmonic ratio of the pencil O.ABCD corresponding to any given ratio of the range is the same function of the sines of the angles as the given ratio is of the corresponding segments.

Thus $\dfrac{\sin AOC . \sin BOD}{\sin AOD . \sin BOC}$ corresponds to $\dfrac{AC.BD}{AD.BC}$;

or, symbolically, $O\{ABCD\}$ corresponds to $\{ABCD\}$.

HARMONIC AND ANHARMONIC DIVISION. 257

To prove that the corresponding anharmonic ratios of the range and pencil are equal.

$$\frac{AC}{BC} = \frac{\triangle AOC}{\triangle BOC} = \frac{OA \cdot OC \sin AOC}{OB \cdot OC \sin BOC} = \frac{OA}{OB} \cdot \frac{\sin AOC}{\sin BOC}.$$

Similarly, $\dfrac{BD}{AD} = \dfrac{OB}{OA} \cdot \dfrac{\sin BOD}{\sin AOD}$,

∴ $\dfrac{AC \cdot BD}{AD \cdot BC} = \dfrac{\sin AOC \cdot \sin BOD}{\sin AOD \cdot \sin BOC}.$

Hence, symbolically
$$\{ABCD\} = O\{ABCD\};$$
and, with necessary formal variations, the anharmonic ratio of a range may be changed for that of the corresponding pencil, and *vice versa*, whenever required to be done.

Cor. 1. Two angles with a common vertex divide each other equianharmonically. (300°)

Cor. 2. If the anharmonic ratio of a pencil is $+1$, two rays coincide, and if -1, the pencil is harmonic. (299°, 4, 5)

Cor. 3. A given range determines an equianharmonic pencil at every vertex, and a given pencil determines an equianharmonic range on every transversal.

Cor. 4. Since the sine of an angle is the same as the sine of its supplement (214°, 1), any ray may be rotated through a straight angle or reversed in direction without affecting the ratio.

Corollaries 2, 3, and 4 are of special importance.

305. *Theorem.*—If three pairs of corresponding rays of two equianharmonic pencils intersect collinearly, the fourth pair intersect upon the line of collinearity.

Proof.—Let
$$O\{ABCD\} = O'\{ABCD'\},$$
and let the pairs of corresponding rays OA and O'A, OB and O'B, OC and O'C intersect in the three collinear points A, B, and C. Let the fourth corresponding rays meet the axis of ABC in D and D' respectively. Then $\{ABCD\} = \{ABCD'\}$, (304°)

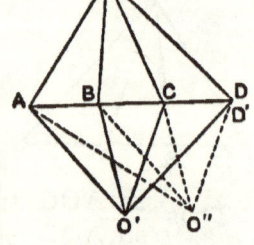

R

$$\therefore \quad \frac{AC \cdot BD}{AD \cdot BC} = \frac{AC \cdot BD'}{AD' \cdot BC'}, \text{ and } \frac{BD}{AD} = \frac{BD'}{AD'},$$

which is possible only when D and D' coincide.

∴ the fourth intersection is upon the axis of A, B, and C, and the four intersections are collinear. *q.e.d.*

Cor. If two of the corresponding rays as OC and O"C become one line, these rays may be considered as intersecting at all points on this line, and however A and B are situated three corresponding pairs of rays necessarily intersect collinearly.

∴ when two equianharmonic pencils have a pair of corresponding rays in common, the remaining rays intersect collinearly.

306°. *Theorem.*—If two equianharmonic ranges have three pairs of corresponding points in perspective, the fourth points are in the same perspective.

Proof.—
$$\{ABCD\} = \{A'B'C'D'\},$$
and A and A', B and B', and C and C' are in perspective at O. Now
$$O\{ABCD\} = O\{A'B'C'D'\},$$
and we have two equianharmonic pencils of which three pairs of corresponding rays meet collinearly at A, B, and C. Therefore OD' and OD meet at D, or D and D' are in perspective at O.

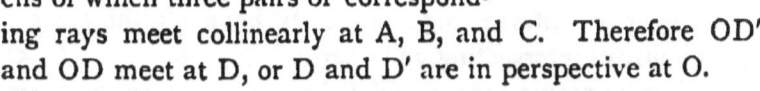

Cor. If two of the corresponding points, as C and C", become coincident, these two points are in perspective at every centre, and hence three corresponding pairs of points are necessarily in perspective.

∴ when two equianharmonic ranges have a pair of corresponding points coincident, the remaining pairs of corresponding points are in perspective.

SECTION II.

HARMONIC RATIO.

307°. Harmonic ratio being a special case of anharmonic ratio (299°, 5), the properties and relations of the latter belong also to the former.

The harmonic properties of a divided segment may accordingly be classified as follows :—

1. The dividing points alternate with the end points of the divided segment.

For this reason harmonic division is symbolized by writing the letters in order of position, as, $\{APBQ\}$, where A and B are the end points of the segment and P and Q the dividing points (301°). A—P—B—Q.

2. The dividing points P and Q divide the segment externally and internally in the same ratio, neglecting sign. (299°, 5)

3. If one segment divides another harmonically, the second also divides the first harmonically. (300°)

4. A harmonic range determines a harmonic pencil at every vertex, and a harmonic pencil determines a harmonic range on every transversal. (304°, Cor. 3)

5. If one or more rays of a harmonic pencil be reversed in direction the pencil remains harmonic. (304°, Cor. 4)

6. Two harmonic pencils which have three pairs of corresponding rays intersecting collinearly have all their corresponding rays intersecting collinearly. (305°)

7. Two harmonic ranges which have three pairs of corresponding points in perspective have all their corresponding points in perspective. (306°)

8. If two harmonic pencils have a corresponding ray from each in common, all their corresponding rays intersect collinearly. (305°, Cor.)

9. If two harmonic ranges have a corresponding point from each in common, all their corresponding points are in perspective. (306°, Cor.)

308°. Let APBQ be a harmonic range. Then
$$AP : PB = AQ : BQ,$$
$$\therefore AP : AQ = AB - AP : AQ - AB.$$
Taking AP, AB, AQ as three magnitudes, we have the statement:—

The first is to the third as the difference between the first and second is to the difference between the second and the third. And this is the definition of three quantities in Harmonic Proportion as given in Arithmetic and Algebra.

Exercises.

1. When three line segments are in harmonic proportion the rectangle on the mean and the sum of the extremes is equal to twice the rectangle on the extremes.
2. The expanded symbol $\{APBQ\} = -1$ gives $AP : AQ = -BP : BQ$. Why the negative sign?
3. Prove from the nature of harmonic division that when P bisects AB, Q is at ∞.
4. Prove that if OP bisects $\angle AOB$ internally OQ bisects it externally; $O\{APBQ\}$ is equal to -1.
5. Trace the changes in the value of the ratio AC : BC as C moves from $-\infty$ to $+\infty$.

309°. In the harmonic range APBQ, P and Q are called conjugate points, and so also are A and B.

Similarly in the harmonic pencil O . APBQ, OP and OQ are conjugate rays, and so also are OA and OB.

Ex. 1. Given three points of a harmonic range to find the fourth.

Let A, P, B be the three given points.

By (259°, Ex. 7) find any point O at which the segments AP and PB subtend equal angles. Draw OQ the external bisector of the $\angle AOB$. Q is the fourth point.

For OP and OQ are internal and external bisectors of the $\angle AOB$. (208°, Cor. 1)

Ex. 2. Given three rays to find a fourth so as to make the pencil harmonic.

Let OA, OP, OB be the three rays. On OA take any two equal distances OD and DE.

Draw DF ∥ to OB, and draw OQ ∥ to EF. OQ is the fourth ray required.

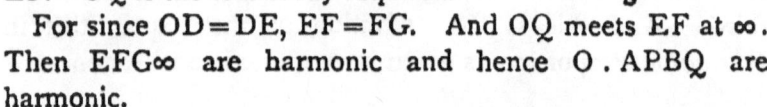

For since OD = DE, EF = FG. And OQ meets EF at ∞. Then EFG∞ are harmonic and hence O . APBQ are harmonic.

Cor. In the symbolic expression for a harmonic ratio a pair of conjugates can be interchanged without destroying the harmonicism.

$$\{APBQ\}=\{BPAQ\}=\{BQAP\}=\{AQBP\},$$

for . {APBQ} gives AP . BQ : AQ . BP = −1,
and {BPAQ} gives BP . AQ : BQ . AP,
and being the reciprocal of the former its value is −1 also.

And similarly for the remaining symbols.

HARMONIC PROPERTIES OF THE TETRAGRAM OR COMPLETE QUADRANGLE.

310°. Let ABCD be a quadrangle, of which AC, BD, and EF are the three diagonals. (247°, Def. 2)

Also let the line EO cut two sides in G and H, and the line FO cut the other two sides in K and L. Then

1. AED is a △ whereof AC, EH, and DB are concurrent lines from the vertices to the opposite sides.

∴ AB . EC . DH = −AH . DC . EB. (251°, b)

Also, AED is a △ and FCB is a transversal.

∴ AB . EC . DF = AF . DC . EB, (250°, a)

262   SYNTHETIC GEOMETRY.

and dividing the former equality by the latter,

$$\frac{DH}{DF} = -\frac{AH}{AF},$$

and AHDF is a harmonic range.

2. Again, $\{AHDF\} = E\{AHDF\} = E\{LOKF\} = E\{BGCF\}$.
   $(307°, 4)$

∴   LOKF and BGCF are harmonic ranges.

3. $O\{AHDF\} = O\{CEDK\} = F\{CEDK\} = F\{GEHO\}$
   $= F\{BEAL\},$ $(307°, 4, 5)$
but $\{CEDK\} = \{DKCE\}$, etc. $(309°, Cor.)$
∴   DKCE, HOGE, ALBE are harmonic ranges.

4. If AC be produced to meet EF in I, AOCI is a harmonic range.

∴ all the lines upon which four points of the figure lie are divided harmonically by the points.

And the points E, F, and O at which four lines concur are vertices of harmonic pencils.

Exercises.

1. A line ∥ to the base of a △ has its points of intersection with the sides connected transversely with the end points of the base. The join of the vertex with the point of intersection of these connectors is a median, and is divided harmonically.

   (Let F go to ∞ in the last figure.)

2. ABC is a △ and BD is an altitude. Through any point O on BD, CO and OA meet the sides in F and E respectively. Show that DE and DF make equal angles with AC.

3. The centres of two circles and their centres of similitude form a harmonic range.

4. In the Fig. of 310° the joins DI, IB, BH, and LD are all divided harmonically.

311°. Let APBQ be a harmonic range and let C be the middle point of AB. Then

$$\frac{AP}{AQ} = -\frac{BP}{BQ} = \frac{PB}{BQ}.$$

That is $\dfrac{CB+CP}{CB+CQ} = \dfrac{CB-CP}{CQ-CB}$,

or $\dfrac{CB+CP}{CB-CP} = \dfrac{CQ+CB}{CQ-CB}$,

whence $\dfrac{CB}{CP} = \dfrac{CQ}{CB}$.

∴ CP . CQ = CB², or P and Q are inverse points to the circle having C as centre and CB as radius.

∴ 1. *The diameter of a circle is divided harmonically by any pair of inverse points.*

And a circle having a pair of conjugates of a harmonic range as end-points of a diameter has the other pair of conjugates as inverse points.

Again, let EF be any secant through P meeting the polar of P in V.

A circle on PV as diameter passes through Q and P, and therefore cuts S orthogonally. (258°)

Hence also the circle S cuts the circle on PV orthogonally, and E and F are inverse points to the circle on PV.

∴ EPFV is a harmonic range.

∴ 2. *A line is cut harmonically by a point, a circle, and the polar of the point with respect to the circle.*

Ex. P, Q are inverse points, and from Q a line is drawn cutting the circle in A and B. The join PB cuts the circle in A'. Then AA' is ⊥ to PQ.

312°. Let P be any point and L its polar with respect to the circle Z. And let PCD and PBA be any two secants. Then

1. PCED and PBFA are harmonic ranges having P a corresponding point in each. Therefore AD, FE, and BC are concurrent. And BC and AD meet on the polar of P. (309°, 9)

2. Again, since {PCED}={PDEC},   (309°, Cor.)
∴ PDEC and PBFA are harmonic ranges having P a corresponding point in each. Therefore DB, EF, and CA are concurrent, and AC and DB meet on the polar of P.

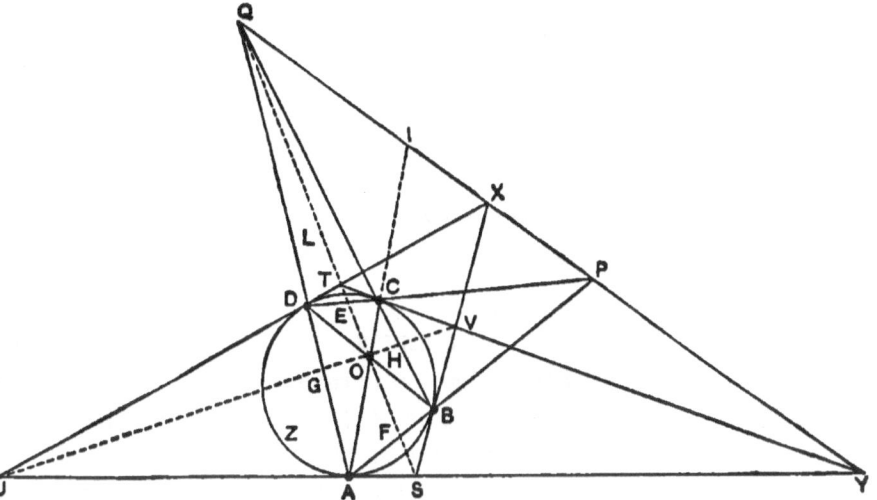

∴ *If from any point two secants be drawn to a circle, the connectors of their points of intersection with the circle meet upon the polar of the first point.*

3. Since O is on the polar of P, P is on the polar of O. But since Q is a point from which secants are drawn satisfying the conditions of 2, Q is on the polar of O.

∴ PQ is the polar of O.

Now ABCD is a concyclic quadrangle whereof AC and BD are internal diagonals and PQ the external diagonal.

∴ *In any concyclic quadrangle the external diagonal is the polar of the point of intersection of the internal diagonals, with respect to the circumcircle.*

4. Since Q is on the polar of P and also on that of O, therefore PO is the polar of Q, and POQ is a triangle self-conjugate with respect to the circle.

5. Let tangents at the points A, B, C, D form the circumscribed quadrangle USVT.

HARMONIC RATIO.

Then S is the pole of AB, and T of DC.

∴ ST is the polar of P, and S and T are points on the line QO.

Similarly U and V are points on the line PO.

But XY is the external diagonal of USVT, and its pole is O, the point of intersection of DB and AC.

∴ X and Y are points on the line PQ.

Hence, *If tangents be drawn at the vertices of a concyclic quadrangle so as to form a circumscribed quadrangle, the internal diagonals of the two quadrangles are concurrent, and their external diagonals are segments of a common line; and the point of concurrence and the line are pole and polar with respect to the circle.*

EXERCISES.

1. UOVP and SOTQ are harmonic ranges.
2. If DB meets the line PQ in R, IOR is a self-conjugate triangle with respect to the circle.
3. To find a circle which shall cut the sides of a given triangle harmonically.
4. QXPY is a harmonic range.

313°. Let S be a circle and A′P′B′Q′ a harmonic range.

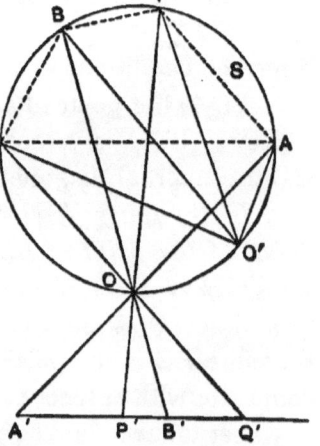

Taking any point O on the circle and through it projecting rectilinearly the points A′P′B′Q′ we obtain the system APBQ, which is called a *harmonic system of points* on the circle.

Now, taking O′ any other point on the circle, O′. APBQ is also harmonic. For

∠AOP = ∠AO′P,
∠POB = ∠PO′B, etc.

∴ *Def.*—Four points on a circle form a harmonic system when their joins with any fifth point on the circle form a harmonic pencil.

Cor. 1. Since $\sin \angle AOP = \dfrac{AP}{d}$, $\sin \angle POB = \dfrac{PB}{d}$, etc., (228°)

∴ (304°), neglecting sign, $AP \cdot BQ = AQ \cdot PB$,

∴ *When four points form a harmonic system on a circle, the rectangles on the opposite sides of the normal quadrangle which they determine are equal.*

Cor. 2. If O comes to A, the ray OA becomes a tangent at A.

∴ *When four points form a harmonic system on a circle, the tangent at any one of them and the chords from the point of contact to the others form a harmonic pencil.*

314°. Let the axis of the harmonic range APBQ be a tangent to the circle S.

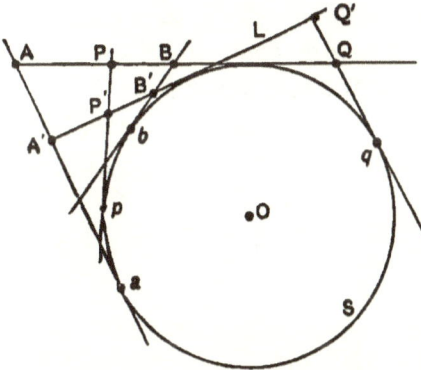

Through A, P, B, and Q draw the tangents A*a*, B*b*, P*p*, and Q*q*.

These four tangents form a harmonic system of tangents to the circle S.

Let L be any other tangent cutting the four tangents of the system in A', P', B', and Q'.

Then, considering A*a*, P*p*, B*b*, etc., as fixed tangents, and A'P'B'Q' as any other tangent.

$\angle AOP = \angle AO'P'$, $\angle POB = \angle P'OB'$, etc., (116°, Ex. 1)

∴ the pencils O . APBQ and O . A'P'B'Q' are both harmonic, and A'P'B'Q' is a harmonic range.

∴ *When four tangents form a harmonic system to a circle, they intersect any other tangent in points which form a harmonic range.*

Cor. 1. If the variable tangent coincides with one of the fixed tangents, the point of contact of the latter becomes one of the points of the range.

∴ *When four tangents form a harmonic system to a circle, each tangent is divided harmonically by its point of contact and its intersections with the other tangents.*

EXERCISES.

1. Tangents are drawn at A, A' the end-points of a diameter, and two points P, B are taken on the tangent through A such that AB=2AP. Through P and B tangents are drawn cutting the tangent at A' in P' and B'. Then 2A'B'=A'P', and AA', PB', and BP' are concurrent.
2. Four points form a harmonic system on a circle. Then the tangents at one pair of conjugates meet upon the secant through the other pair.
3. If four tangents form a harmonic system to a circle, the point of intersection of a pair of conjugate tangents lies on the chord of contact of the remaining pair.
4. If four points form a harmonic range, their polars with respect to any circle form a harmonic pencil; and conversely.

SECTION III.

OF ANHARMONIC PROPERTIES.

315°. Let A, E, C and D, B, F be two sets of three collinear points having their axes meeting in some point R.

Join the points alternately, as ABCDEFA. Then AB and DE, BC and EF, CD and FA meet in P, Q, O. To show that these points are collinear.

$$\begin{aligned}
O\{ECQF\} &= C\{EOQF\} &&\text{(referred to axis EF)}\\
&= C\{RDBF\} &&\text{(referred to axis DR)}\\
&= A\{RDBF\}\\
&= A\{EDPF\} &&\text{(referred to axis DE)}\\
&= O\{EDPF\}\\
&= O\{ECPF\}. &&\text{(by reversing rays, etc.)}
\end{aligned}$$

∴ the pencils O.ECQF and O.ECPF are equianharmonic, and having three rays in common the fourth rays must be in common, *i.e.*, they can differ only by a straight angle, and therefore O, P, Q are collinear.

(Being the first application of anharmonic ratios the work is very much expanded.)

∴ *If six lines taken in order intersect alternately in two sets of three collinear points, they intersect in a third set of three collinear points.*

Cor. 1. ABC and DEF are two triangles, whereof each has one vertex lying upon a side of the other.

If AB and DE are taken as corresponding sides, A and F are non-corresponding vertices. But, if AB and EF are taken as corresponding sides, A and D are non-corresponding vertices.

Hence the intersections of AB and EF, of ED and CB, and of AD and CF are collinear.

∴ If two triangles have each a vertex lying upon a side of the other, the remaining sides and the joins of the remaining non-corresponding vertices intersect collinearly.

Cor. 2. Joining AD, BE, CF, ADBE, EBFC, and ADFC are quadrangles, and P, Q, O are respectively the points of intersection of their internal diagonals.

∴ if a quadrangle be divided into two quadrangles, the points of intersection of the internal diagonals of the three quadrangles are collinear.

316°. Let A, A', B, B', C, C' be six points lying two by two on two sets of three concurrent lines, which meet at P and Q. Then the points lie upon a third set of three concurrent lines meeting at O.

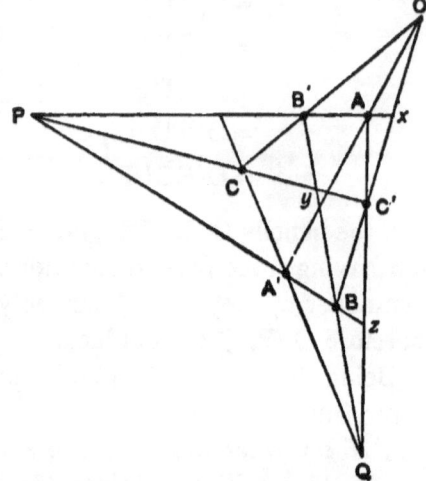

We are to prove that AO and AA' are in line.

$$A\{OxC'B\} = B'\{OxC'Q\}$$
$$= B'\{CPC'y\}$$
$$= Q\{A'PzB\}$$
$$= A\{A'xC'B\}.$$

∴ the pencils A.OxC'B and A.A'xC'B are equianharmonic, and have three corresponding rays in common. Therefore AO and AA' are in line.

Cor. ABC and A'B'C' are two △s which are in perspective at both P and Q, and we have shown that they are in perspective at O also.

As there is an axis of perspective corresponding to each centre, the joins of the six points, accented letters being taken together and unaccented together, taken in every order intersect in three sets of three collinear points.

EXERCISES.

1. If two △s have their sides intersecting collinearly, their corresponding vertices connect concurrently.
2. The converse of Ex. 1.
3. Three equianharmonic ranges ABCD, A'B'C'D', and PQRS have their axes concurrent at Y, and AA', BB', CC', DD' concurrent at X. Then the two groups of joins AP, BQ, CR, DS, and A'P, B'Q, C'R, D'S are

270   SYNTHETIC GEOMETRY.

concurrent at two points O and O' which are collinear with X.

4. From Ex. 3 show that if a variable △ has its sides passing through three fixed points, and two of its vertices lying upon fixed lines, its third vertex lies upon a fixed line concurrent with the other two.

5. If a variable △ has its vertices lying on three fixed lines and two of its sides passing through fixed points, its third side passes through a fixed point collinear with the other two.

317°. The range ABCD is transferred to the circle S by rectilinear projection through any point O on the circle. Then A'B'C'D' is a system of points on the circle which is equianharmonic with the range ABCD.

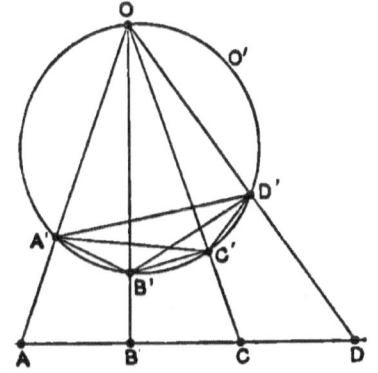

1. If O' be any other point on the circle, the
$$\angle A'OB' = \angle A'O'B',$$
$$\angle B'OC' = \angle B'O'C', \text{ etc.,}$$
and the two pencils O . A'B'C'D' and O' . A'B'C'D' are equianharmonic.

∴ four points on a circle subtend equianharmonic pencils at all fifth points on the circle.

2. Since
$$\sin A'OB' = \frac{A'B'}{d},$$
$$\sin C'OD' = \frac{C'D'}{d}, \text{ etc.,}$$

and ∵   AB . CD + BC . AD = AC . BD,   (233°)

∴   A'B' . C'D' + B'C' . A'D' = A'C' . B'D';

which is an extension of Ptolemy's theorem to a concyclic quadrangle.   (205°)

OF ANHARMONIC PROPERTIES. 271

3. If the range ABCD is inverted with O as the centre of inversion, the axis of the range inverts into a circle S through O, and A, B, C, D invert into A', B', C', and D' respectively.

Hence, in general, anharmonic relations are unchanged by inversion, a range becoming an equianharmonic system on a circle, and under certain conditions *vice versa*.

4. In the inversion of 3, A and A', B and B', etc., are pairs of inverse points.

$$\therefore \quad OA \cdot OA' = OB \cdot OB'$$
$$= OC \cdot OC' = OD \cdot OD',$$

and the $\triangle$s OAB and OB'A', OAC and OC'A', etc., are similar in pairs.

And if P be the $\perp$ from O to AD, and P_1, P_2, P_3, and P_4 be the $\perp$s from O to A'B', B'C', C'D', and D'A', we have

$$\frac{AB}{P} = \frac{A'B'}{P_1}, \quad \frac{BC}{P} = \frac{B'C'}{P_2},$$

$$\frac{CD}{P} = \frac{C'D'}{P_3}, \quad \frac{DA}{P} = \frac{D'A'}{P_4},$$

$$\therefore \quad \frac{AB + BC + CD + DA}{P} = \frac{A'B'}{P_1} + \frac{B'C'}{P_2} + \frac{C'D'}{P_3} + \frac{D'A'}{P_4}.$$

But (232°) $\quad AB + BC + CD + DA = 0,$

$$\therefore \quad \frac{A'B'}{P_1} + \frac{B'C'}{P_2} + \frac{C'D'}{P_3} + \frac{D'A'}{P_4} = 0,$$

And since the same principle applies to a range of any number of points,

$\therefore$ in any concyclic polygon, if each side be divided by the perpendicular upon it from any fixed point on the circle, the sum of the quotients is zero.

In the preceding theorem, as there is no criterion by which we can distinguish any side as being negative, some of the perpendiculars must be negative.

Of the perpendiculars one falls externally upon its side of

272 SYNTHETIC GEOMETRY.

the polygon and all the others fall internally. Therefore the theorem may be stated :—

If ⊥s be drawn from any point on a circle to the sides of an inscribed polygon, the ratio of the side, upon which the ⊥ falls externally, to its ⊥ is equal to the sum of the ratios of the remaining sides to their ⊥s.

318°. *Theorem.*—If two circles be inverted the ratio of the square on their common tangent to the rectangle on their diameters is unchanged.

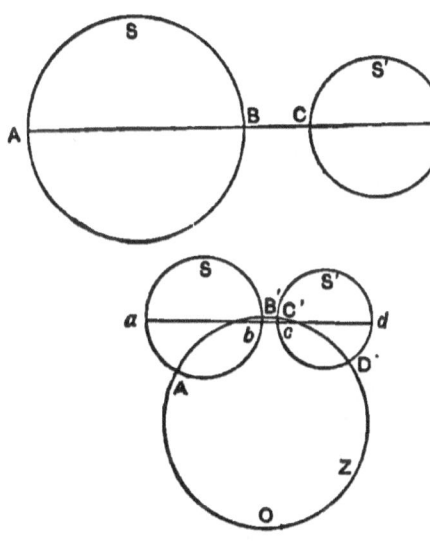

Let S, S' be the circles and AD be the common centre line, and let the circles s and s' and the circle Z be their inverses respectively.

Then Z cuts s and s' orthogonally, and O{ABCD}=O{A'B'C'D'}.

But if *abcd* be the common centre line of s and s', aA', bB', cC', and dD' are concurrent at O.   (265°, Ex. 1)

∴   O{*abcd*}=O{A'B'C'D'}
           =O{ABCD},

∴   {*abcd*}={ABCD},

or $\dfrac{AC.BD}{AB.CD} = \dfrac{ac.bd}{ab.cd}.$

But AC.BD = the square on the common direct tangent to S and S', and $ac.bd$ = the square of the corresponding tangent to s and s'.   (179°, Ex. 2)

And AB.CD and $ab.cd$ are the products of the diameters respectively.

And the theorem is proved.

Cor. 1. Writing the symbolic expressions {ABCD} and {abcd} in another form, we have
$$\frac{AD \cdot BC}{AB \cdot CD} = \frac{ad \cdot bc}{ab \cdot cd}.$$
And AD.BC and $ad.bc$ are equal to the squares on the transverse common tangents respectively.

Cor. 2. If four circles S_1, S_2, S_3, S_4 touch a line at the points A, B, C, D, and the system be inverted, we have four circles s_1, s_2, s_3, s_4, which touch a circle Z through the centre of inversion.

Now let d_1, d_2, d_3, d_4 be the diameters of S_1, S_2, etc., and let δ_1, δ_2, δ_3, δ_4 be the diameters of s_1, s_2, etc., and let t_{12} be the common tangent to s_1 and s_2, t_{34} be that to s_3 and s_4, etc.

Then AB, etc., are common tangents to S_1 and S_2, etc., and
$$AB \cdot CD + BC \cdot AD + CA \cdot BD = 0. \qquad (233°)$$
And
$$\frac{AB^2}{d_1 d_2} = \frac{t_{12}^2}{\delta_1 \delta_2},$$
$$\frac{CD^2}{d_3 d_4} = \frac{t_{34}^2}{\delta_3 \delta_4},$$
$$\therefore \frac{AB \cdot CD}{\sqrt{d_1 d_2 d_3 d_4}} = \frac{t_{12} t_{34}}{\sqrt{\delta_1 \delta_2 \delta_3 \delta_4}},$$
and similar equalities for the remaining terms,
$$\therefore \quad t_{12}t_{34} + t_{23}t_{14} + t_{31}t_{24} = 0.$$

This theorem, which is due to Dr. Casey, is an extension of Ptolemy's theorem. For, if the circles become point-circles, the points form the vertices of a concyclic quadrangle and the tangents form its sides and diagonals.

If we take the incircle and the three excircles of a triangle as the four circles, and the sides of the triangle as tangents, we obtain by the help of Ex. 1, 135°, $b^2 - c^2 + c^2 - a^2 + a^2 - b^2$ as the equivalent for $t_{12}t_{34}$+etc.; and as this expression is identically zero, the four circles given can all be touched by a fifth circle.

319°. Let A, B, C, D, E, F be six points on a circle so connected as to form a hexagram, *i.e.*, such that each point is connected with two others.

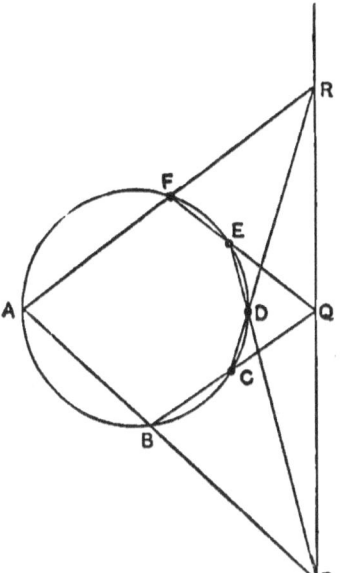

Let the opposite sides AB, DE meet in P; BC, EF in Q; and CD, FA in R.

To prove that P, Q, and R are collinear.

$$Q\{BDER\} = Q\{CDER\}$$
$$= F\{CDEA\}$$
$$= B\{CDEP\}$$
$$= Q\{BDEP\},$$

∴ QR and QP are in line.

∴ if a hexagram have its vertices concyclic, the points of intersection of its opposite sides in pairs are collinear.

Def.—The line of collinearity is called the *pascal* of the hexagram, after the famous Pascal who discovered the theorem, and the theorem itself is known as *Pascal's theorem*.

Cor. 1. The six points may be connected in 5 × 4 × 3 × 2 or 120 different ways. For, starting at A, we have five choices for our first connection. It having been fixed upon, we have four for the next, and so on to the last. But one-half of the hexagrams so described will be the other half described by going around the figure in an opposite direction. Hence, six points on a circle can be connected so as to form 60 different hexagrams. Each of these has its own pascal, and there are thus 60 pascal lines in all.

When the connections are made in consecutive order about the circle the pascal of the hexagram so formed falls without the circle; but if any other order of connection is taken, the pascal may cut the circle.

Cor 2. In the hexagram in the figure, the pascal is the line through P, Q, R cutting the circle in H and K. Now

$$C\{KFBD\} = C\{KFQR\}$$
$$= F\{KCQR\}$$
$$= F\{KCEA\},$$
$$\therefore \quad \{KFBD\} = \{KCEA\},$$

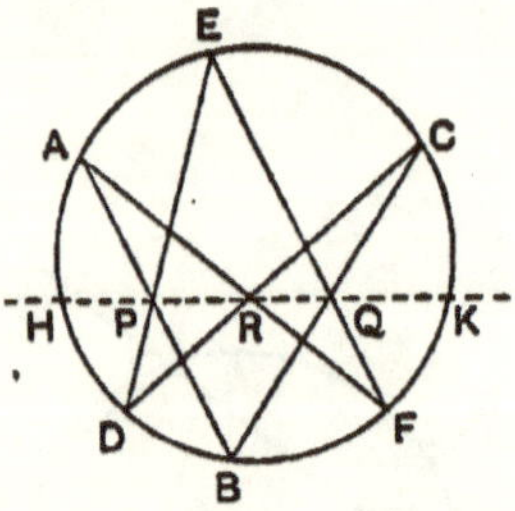

and K is a common point to two equianharmonic systems on the circle. So also is H.

These points are important in the theory of homographic systems.

Cor. 3. Let 1, 2, 3, 4, 5, 6 denote six points taken consecutively upon a circle. Then any particular hexagram is denoted by writing the order in which the points are connected, as for example, 2461352.

In the hexagram 246135 the pairs of opposite sides are 24 and 13, 46 and 35, 61 and 52, and the pascal passes through their intersections.

Now taking the four hexagrams

$$246135, \ \overline{245136}, \ \overline{246315}, \ \overline{245316},$$

the pascal of each passes through the intersection of the connector of 2 and 4 with the connector of 1 and 3. Hence the pascals of these four hexagrams have a common point.

It is readily seen that inverting the order of 2 and 4 gives hexagrams which are only those already written taken in an inverted order.

$\therefore$ the pascals exist in concurrent groups of four, meeting at fifteen points which are intersections of connectors.

Cor. 4. In the hexagram 1352461 consider the two triangles formed by the sides 13, $\overline{52}$, 46 and 35, $\overline{24}$, $\overline{61}$. The sides 13 and $\overline{24}$, $\overline{35}$ and $\overline{46}$, $\overline{52}$ and $\overline{61}$ intersect on the pascal of 1352461, and therefore intersect collinearly.

Hence the vertices of these triangles connect concurrently, *i.e.*, the line through the intersection of 35 and $\overline{61}$ and the intersection of $\overline{52}$ and $\overline{46}$, the line through the intersection of

35 and $\overline{24}$ and the intersection of $\overline{13}$ and $\overline{46}$, and the line through the intersections of $\overline{24}$ and $\overline{61}$ and the intersection of $\overline{13}$ and $\overline{25}$ are concurrent.

But the first of these lines is the pascal of the hexagram 1643521, the second is the pascal of the hexagram 3564213, and the third is the pascal of the hexagram 4256134.

∴ the pascals exist in concurrent groups of three, meeting at 20 points distinct from the 15 points already mentioned.

Cor. 5. If two vertices of the hexagram coincide, the figure becomes a pentagram, and the missing side becomes a tangent.

∴ if a pentagram be inscribed in a circle and a tangent at any vertex meet the opposite side, the point of intersection and the points where the sides about that vertex meet the remaining sides are collinear.

Ex. 1. The tangents at opposite vertices of a concyclic quadrangle intersect upon the external diagonal of the quadrangle.

Ex. 2. ABCD is a concyclic quadrangle. AB and CD meet at E, the tangent at A meets BC at G, and the tangent at B meets AD at F. Then E, F, G are collinear.

320°. Let six tangents denoted by the numbers 1, 2, 3, 4, 5, and 6 touch a circle in A, B, C, D, E, and F. And let the points of intersection of the tangents be denoted by 12, 23, 34, etc.

Then the tangents form a hexagram about the circle.

Now, 12 is the pole of AB, and 45 is the pole of ED. Therefore the

OF ANHARMONIC PROPERTIES. 277

line 12.45 is the polar of the point of intersection of AB and ED.

Similarly the line 23.56 is the polar of the intersection of BC and EF, and the line 34.61 is the polar of the intersection of CD and FA.

But since ABCDEF is a hexagram in the circle, these three intersections are collinear. (319°)

∴ the lines 12.45, 23.56, and 34.61 are concurrent at O.

And hence the hexagram formed by any six tangents to a circle has its opposite vertices connecting concurrently.

Def.—The point of concurrence is the *Brianchon* point, and the theorem is known as *Brianchon's theorem*.

Cor. 1. As the six tangents can be taken in any order to form the hexagram, there are 60 different hexagrams each having its own Brianchon point.

Now take, as example, the hexagram formed by the lines $\overline{123456}$ taken in order.

The connectors are 12.45, 23.56, 34.61, and these give the point O.

But the hexagrams $\overline{126453}$, $\overline{123546}$, and $\overline{126543}$ all have one connector in common with $\overline{123456}$, namely, that which passes through 12 and 45. Hence the Brianchon points of these four hexagrams lie upon one connector.

∴ the 60 Brianchon points lie in collinear groups of four upon 15 connectors of the points of intersection of the tangents.

Cor. 2. Consider the triangles 12.56.34 and 45.23.61. These have their vertices connecting concurrently, and therefore they have their sides intersecting collinearly.

But the point of intersection of the sides 61.23 and 56.34 is the Brianchon point of the hexagram formed by the six lines 234165 taken in order; and similar relations apply to the other points of intersection.

Hence the 60 Brianchon points lie in collinear groups of three upon axes which are not diagonals of the figure.

Cor. 3. Let two of the tangents become coincident.

Their point of intersection is then their common point of contact, and the hexagram becomes a pentagram.

∴ in any pentagram circumscribed to a circle the join of a point of contact with the opposite vertex is concurrent with the joins of the remaining vertices in pairs.

Ex. 1. In any quadrangle circumscribed to a circle, the diagonals and the chords of contact are concurrent.

Ex. 2. In any quadrangle circumscribed to a circle, the lines joining any two vertices to the two points of contact adjacent to a third vertex intersect on the join of the third and the remaining vertex.

SECTION IV.

OF POLAR RECIPROCALS AND RECIPROCATION.

321°. The relation of pole and polar has already been explained and somewhat elucidated in Part IV., Section V.

It was there explained that when a figure consists of any number of points, and their connecting lines, another figure of the same species may be obtained by taking the poles of the connectors of the first figure as points, and the polars of the points in the first figure as connecting lines to form the second.

And as the first figure may be reobtained from the second in the same way as the second is obtained from the first, the figures are said to be *polar reciprocals* of one another, as being connected by a kind of reciprocal relation. The word reciprocal in this connection has not the same meaning as in 184°, Def.

The process by which we pass from a figure to its polar reciprocal is called *polar reciprocation* or simply *reciprocation*.

OF POLAR RECIPROCALS AND RECIPROCATION. 279

322°. Reciprocation is effected with respect to a circle either expressed or implied. The radius and centre of this reciprocating circle are quite arbitrary, and usually no account need be taken of the radius. Certain problems in reciprocation, however, have reference to the centre of reciprocation, although the position of that centre may generally be assumed at pleasure.

From the nature of reciprocation we obtain at once the following statements :—

1. A point reciprocates into a line and a line into a point. And hence a figure consisting of points and lines reciprocates into one consisting of lines and points.

2. Every rectilinear figure consisting of more than a single line reciprocates into a rectilinear figure.

3. The centre of reciprocation reciprocates into the line at ∞, and a centre-line of the circle of reciprocation reciprocates into a point at ∞ in a direction orthogonal to that of the centre-line.

4. A range of points reciprocates into a pencil of lines, and the axis of the range into the vertex of the pencil. And similarly, a pencil of lines reciprocates into a range of points, and the vertex of the pencil into the axis of the range.

323°. Let O.LMNK be a pencil of four, and C be the centre of reciprocation. Draw the perpendiculars Cl' on L, Cm' on M, Cn' on N, and Ck' on K.

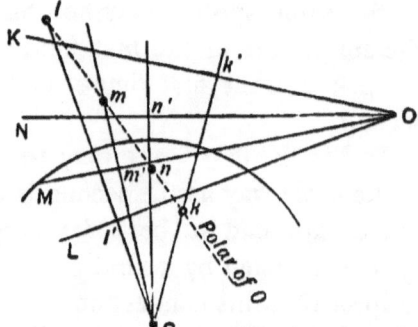

The poles of L, M, N, K lie respectively on these perpendiculars, forming a range of points as l, m, n, k. Then

1. Evidently the
$$\angle LOM = \angle l'Cm, \quad \angle MON = \angle mCn, \text{ etc.}$$
∴ the angle between two lines is equal to that subtended at

the centre of reciprocation by the poles of the lines; and the angle subtended at the centre of reciprocation by two points is equal to the angle between the polars of the points.

2. Any pencil of four is equianharmonic with its polar reciprocal range. And hence anharmonic or harmonic relations are not altered by reciprocation.

Def.—Points are said to be perpendicular to one another when their joins with the centre of reciprocation are at right angles. In such a case the polars of the points are perpendicular to one another.

324°. In many cases, and especially in rectilinear figures, the passing from a theorem to its polar reciprocal is quite a mechanical process, involving nothing more than an intelligent and consistent change in certain words in the statement of the theorem.

In all such cases the truth of either theorem follows from that of its polar reciprocal as a matter of necessity.

Take as example the theorem of 88°, "The three altitudes of a $\triangle$ are concurrent."

To get its polar reciprocal put it in the following form, where the theorem and its polar reciprocal are given in alternate lines:—

The three $\begin{cases} \text{lines through the vertices} \\ \text{points on the sides} \end{cases}$ of a $\triangle$ perpendicular to the opposite $\begin{cases} \text{sides} \\ \text{vertices} \end{cases}$ are $\begin{cases} \text{concurrent.} \\ \text{collinear.} \end{cases}$

To get a point $\perp$ to a vertex we connect the vertex to the centre of reciprocation, and through this centre draw a line $\perp$ to the connector. The point required lies somewhere on this line. (323°, Def.)

And as the centre may be any point, we may state the polar reciprocal thus:—

"The lines through any point perpendicular to the joins of

OF POLAR RECIPROCALS AND RECIPROCATION. 281

that point with the vertices of a triangle intersect the opposite sides of the triangle collinearly." (252°, Ex. 8)

325°. Consider any two $\triangle$s. These reciprocate into two $\triangle$s; vertices giving sides, and sides, vertices.

If the original $\triangle$s are in perspective their vertices connect concurrently. But in reciprocation the vertices become sides and the point of concurrence becomes a line of collinearity. Hence the polar reciprocals of these $\triangle$s have their sides intersecting collinearly and are in perspective.

∴ $\triangle$s in perspective reciprocate into $\triangle$s in perspective.

But any three concurrent lines through the vertices of a $\triangle$ intersect the opposite sides in points which form the vertices of a new $\triangle$ in perspective with the former.

Hence all cases of three concurrent lines passing through the vertices of a $\triangle$ reciprocate into $\triangle$s in perspective with the original. Such are the cases of the concurrence of the three medians, the concurrence of the three altitudes, of the three bisectors of the angles, etc.

326°. The complete harmonic properties of the tetragram may be expressed in the two following theorems, which are given in alternate lines, and are polar reciprocals to one another:—

Four $\begin{Bmatrix}\text{lines}\\\text{points}\end{Bmatrix}$ determine by their $\begin{Bmatrix}\text{intersections}\\\text{connectors}\end{Bmatrix}$ six $\begin{Bmatrix}\text{points,}\\\text{lines,}\end{Bmatrix}$ and the $\begin{Bmatrix}\text{connectors}\\\text{intersections}\end{Bmatrix}$ of these by their $\begin{Bmatrix}\text{intersections}\\\text{connectors}\end{Bmatrix}$ determine three new $\begin{Bmatrix}\text{points.}\\\text{lines.}\end{Bmatrix}$ The $\begin{Bmatrix}\text{connectors}\\\text{intersections}\end{Bmatrix}$ of any of the three new $\begin{Bmatrix}\text{points}\\\text{lines}\end{Bmatrix}$ with the original six $\begin{Bmatrix}\text{points}\\\text{lines}\end{Bmatrix}$ form a harmonic $\begin{Bmatrix}\text{pencil.}\\\text{range.}\end{Bmatrix}$

Other polar reciprocal theorems, which have been already

given, are Pascal's and Brianchon's theorems with all their corollaries, the theorems of Arts. 313° and 314°, of Arts. 315° and 316°, etc.

The circle, when reciprocated with respect to any centre of reciprocation not coincident with its own centre, gives rise to a curve of the same species as the circle, *i.e.*, a conic section, and many properties belonging to the circle, and particularly those which are unaltered by reciprocation, become properties of the general curve.

These generalized properties cannot be readily understood without some preliminary knowledge of the conic sections.

SECTION V.

HOMOGRAPHY AND INVOLUTION.

327°. Let A, B and A', B' be fixed points on two lines, and let P and P' be variable points, one on each line which so move as to preserve the relation

$$\frac{AP}{BP} = k \cdot \frac{A'P'}{B'P'},$$

where k is any constant; and let C, C'; D, D'; E, E', etc., be simultaneous positions of P and P'.

Then the points A, B, C, D, E, etc., and A', B', C', D', E', etc., divide homographically the lines upon which they lie.

$$\therefore \quad \frac{AC}{BC} = k \cdot \frac{A'C'}{B'C'} \text{ and } \frac{AD}{BD} = k \cdot \frac{A'D'}{B'D'},$$

$$\therefore \quad \frac{AC \cdot BD}{AD \cdot BC} = \frac{A'C' \cdot B'D'}{A'D' \cdot B'C'},$$

or $\{ABCD\} = \{A'B'C'D'\}$.

Similarly, $\{ABCE\} = \{A'B'C'E'\}$, $\{BCDE\} = \{B'C'D'E'\}$, etc.

HOMOGRAPHY AND INVOLUTION.

Evidently for each position of P, P' can have only one position, and conversely, and hence the points of division on the two axes correspond in unique pairs.

∴ two lines are divided homographically by two sets of points when to each point on one corresponds one and only one point on the other, and when any four points on one line and their four correspondents on the other form equianharmonic ranges.

Cor. 1. If the systems of points be joined to any vertices O and O', the pencils O.ABCD... and O'.A'B'C'D'... are evidently homographic, and cut all transversals in homographic ranges.

Cor. 2. The results of Arts. 304°, Cors. 3 and 4, and of Arts. 305° and 306° and their corollaries are readily extended to homographic ranges and pencils.

The following examples of homographic division are given.

Ex. 1. A line rotating about a fixed point in it cuts any two lines homographically.

Ex. 2. A variable point confined to a given line determines two homographic pencils at any two fixed points.

Ex. 3. A system of *c.p.*-circles determines two homographic ranges upon any line cutting the system.

Consider any two of the circles, let P, Q be the common points, and let the line L cut one of the circles in A and A' and the other in B and B'. Then the $\angle PBB' = \angle PQB'$, and $\angle PAB' = \angle PQA'$. ∴ $\angle APB = \angle A'QB'$.

Hence the segment BA subtends the same angle at P as the segment at B'A' does at Q. And similarly for all the segments made in the other circles.

Ex. 4. A system of *l.p.*-circles determines two homographic ranges upon every line cutting the system.

DOUBLE POINTS OF HOMOGRAPHIC SYSTEMS.

328°. Let ABCD... and A'B'C'D'... be two homographic ranges on a common axis. If any two correspondents from the two ranges become coincident the point of coincidence is a *double point* of the system.

If A and A' were thus coincident we would have the relations $\{ABCE\}=\{AB'C'E'\}$, etc.

Thus a double point is a common constituent of two equianharmonic ranges, of which the remaining constituents are correspondents from two homographic systems upon a common axis.

ABC and A'B'C' being fixed, let D and D' be two variable correspondents of the doubly homographic system.

Then $\{ABCD\}=\{A'B'C'D'\}$,

whence $\dfrac{BD \cdot A'D'}{AD \cdot B'D'} = \dfrac{BC \cdot A'C'}{AC \cdot B'C'} = \dfrac{p}{q}$, say.

Now taking O, an arbitrary point on the axis, let
$OD=x$, $OD'=x'$, $OA=a$, $OA'=a'$, $OB=b$, $OB'=b'$.
Then $BD=x-b$, $B'D'=x'-b'$, $AD=x-a$, $A'D'=x'-a'$,

$\therefore \dfrac{(x-b)(x'-a')}{(x-a)(x'-b')} = \dfrac{p}{q},$

which reduces to the form
$$xx' + Px + Qx' + R = 0.$$
When D and D' become coincident x' becomes equal to x and we have a quadratic from which to determine x, *i.e.*, the positions of D and D' when uniting to form a double point.

Hence every doubly homographic system has two double points which are both real or both imaginary, and of which both may be finite, or one or both may be at infinity.

Evidently there cannot be more than two double points, for since such points belong to two systems, three double points would require the coincidence of three pairs of correspondents, and hence of all. (306°)

329°. If D be one of the double points of a doubly homographic system, $\dfrac{DB \cdot DA'}{DA \cdot DB'} = \dfrac{CB \cdot C'A'}{C'B' \cdot CA} = \dfrac{P}{Q}$, say.

Now DB.DA' and DA.DB' are respectively equal to the squares on tangents from D to any circles passing through B, A' and B', A.

But the locus of a point from which tangents to two given circles are in a constant ratio is a circle co-axal with both.

(275°, Cor. 5)

Hence the following construction for finding the double points.

Through A, B' and A', B draw any two circles so as to intersect in two points U and V, and through these points of intersection pass the circle S'', so as to be the locus of a point from which tangents to the circles S and S' are in the given ratio $\sqrt{P} : \sqrt{Q}$.

The circle S'' cuts the axis in D, D, which are the required double points.

Evidently, instead of A, B' and A', B we may take any pairs of non-corresponding points, as A, C' and A', C; or B, C' and B', C. The given ratio $\sqrt{P} : \sqrt{Q}$ is different, however, for each different grouping of the points.

Cor. 1. When $P = Q$, *i.e.*, when $\dfrac{AC}{BC} = \dfrac{A'C'}{B'C'}$, the circle S'' takes its limiting form of a line and cuts the axis at one finite point or at none.

In this case both double points may be at ∞ or only one of them.

Cor. 2. If any disposition of the constituents of the system causes the circle S'' to lie wholly upon one side of the axis, the double points for that disposition become imaginary.

286 SYNTHETIC GEOMETRY.

330°. Let L be the axis of a doubly homographic system. Through any point O on the circle S transfer the system, by rectilinear projection, to the circle. Then ABC..., A'B'C'... form a doubly homographic system on the circle.

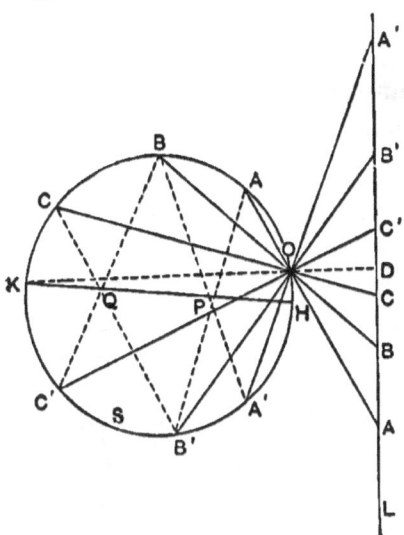

Now, by connecting any two pairs of non-correspondents A, B' and A', B ; B, C' and B', C ; C, A' and C', A, we obtain the pascal line KH which cuts the circle in two points such that

$$\{KABC\} = \{KA'B'C'\}.$$

(319°, Cor. 2)

Hence H and K are double points to the system on the circle. And by transferring K and H back through the point O to the axis L, we obtain the double points D, D of the doubly homographic range.

Cor. 1. When the pascal falls without the circle, the double points are imaginary.

Cor. 2. When one of the joins, KO or HO, is ∥ to L, one of the double points is at ∞.

Cor. 3. If the system upon the circle with its double points H and K be projected rectilinearly through any point on the circle upon any axis M, it is evident that the projected system is a doubly homographic one with its double points.

Cor. 4. Cor. 3 suggests a convenient method of finding the double points of a given axial system.

Instead of employing a circle lying without the axis, employ the axis as a centre-line and pass the circle through any pair of non-correspondents.

Then from any convenient point on the circle transfer the remaining points, find the pascal, and proceed as before.

331°. The following are examples of the application of the double points of doubly homographic systems to the solution of problems.

Ex. 1. Given two non-parallel lines, a point, and a third line. To place between the non-parallels a segment which shall subtend a given angle at the given point, and be parallel to the third line.

Let L and M be the non-parallel lines, and let N be the third line, and O be the given point.

We are to place a segment between L and M, so as to subtend a given angle at O and be ∥ to N.

On L take any three points A, B, C, and join OA, OB, OC. Draw Aa, Bb, Cc all ∥ to N, and draw Oa', Ob', Oc' so as to make the angles AOa', BOb', COc' each equal to the given angle.

Now, if with this construction a coincided with a', or b with b', or c with c', the problem would be solved.

But, if we take a fourth point D, we have
$$O\{ABCD\} = \{ABCD\} = \{abcd\} = \{a'b'c'd'\}.$$
∴ $abcd$ and $a'b'c'd'$ are two homographic systems upon the same axis. Hence the double points of the system give the solutions required.

Ex. 2. Within a given △ to inscribe a △ whose sides shall be parallel to three given lines.

Ex. 3. Within a given △ to inscribe a △ whose sides may pass through three given points.

Ex. 4. To describe a △ such that its sides shall pass through three given points and its vertices lie upon three given lines.

SYSTEMS IN INVOLUTION.

332°. If A, A', B, B' are four points on a common axis, whereof A and A', as also B and B', are correspondents, a point O can always be found upon the axis such that

$$OA \cdot OA' = OB \cdot OB'.$$

This point O is evidently the centre of the circle to which A and A', and also B and B', are pairs of inverse points, and is consequently found by 257°.

Now, let P, P' be a pair of variable conjugate points which so move as to preserve the relation

$$OP \cdot OP' = OA \cdot OA' = OB \cdot OB'.$$

Then P and P' by their varying positions on the axis determine a double system of points C, C', D, D', E, E', etc., conjugates in pairs, so that

$$OA \cdot OA' = OC \cdot OC' = OD \cdot OD' = OE \cdot OE' = \text{etc.}$$

Such a system of points is said to be in *involution*, and O is called the centre of the involution.

When both constituents of any one conjugate pair lie upon the same side of the centre, the two constituents of every conjugate pair lie upon the same side of the centre, since the product must have the same sign in every case.

With such a disposition of the points the circle to which conjugates are inverse points is real and cuts the axis in two points F and F'. At these points variable conjugates meet and become coincident.

Hence the points F, F' are the double points or *foci* of the system.

From Art. 311°, 1, FF' is divided harmonically by every pair of conjugate points, so that

FAF'A', FBF'B', etc., are all harmonic ranges.

When the constituents of any pair of conjugate points lie upon opposite sides of the centre, the foci are imaginary.

HOMOGRAPHY AND INVOLUTION. 289

333°. Let A, A', B, B', C, C' be six points in involution, and let O be the centre.

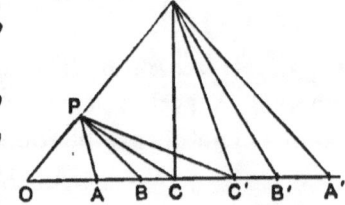

Draw any line OPQ through O, and take P and Q so that
$$OP \cdot OQ = OA \cdot OA',$$
and join PA, PB, PC, and PC', and also QA', QB', QC', and QC.

Then, ∵ OA . OA' = OP . OQ,
∴ A, P, Q, A' are concyclic. ∴ ∠OPA = ∠OA'Q.
Similarly, B, P, Q, B' are concyclic, and ∠OPB = ∠OB'Q, etc.
∴ ∠APB = ∠A'QB'.
Similarly, ∠BPC = ∠B'QC', ∠CPC' = ∠C'QC, etc.
Hence the pencils P(ABCC') and Q(A'B'C'C) are equianharmonic, or {ABCC'} = {A'B'C'C}.
Hence also {ABB'C} = {A'B'BC'}, {AA'BC} = {A'AB'C'}.
And any one of these relations expresses the condition that the six points symbolized may be in involution.

334°. As involution is only a species of homography, the relations constantly existing between homographic ranges and their corresponding pencils, hold also for ranges and pencils in involution. Hence

1. Every range in involution determines a pencil in involution at every vertex, and conversely.

2. If a range in involution be projected rectilinearly through any point on a circle it determines a system in involution on the circle, and conversely.

Ex. The three pairs of opposite connectors of any four points cut any line in a six-point involution.

A, B, C, D are the four points, and P, P' the line cut by the six connectors CD, DA, AC, CB, BD, and AB. Then
$$D\{PQRR'\} = D\{CARB\}$$
$$= B\{CARD\}$$
$$= B\{Q'P'RR'\} = \{P'Q'R'R\}, \qquad (302°)$$
T

∴ $\{PQRR'\} = \{P'Q'R'R\}$,

and the six points are in involution.

Cor. 1. The centre O of the involution is the radical centre of any three circles through PP', QQ', and RR'; and the three circles on the three segments PP', QQ', and RR' as diameters are co-axal.

When the order of PQR is opposite that of P'Q'R' as in the figure, and the centre O lies outside the points, the co-axal circles are of the *l.p.*-species, and when the two triads of points have the same order, the co-axal circles are of the *c.p.*-species.

Cor. 2. Considering ABC as a triangle and AD, BD, CD three lines through its vertices at D, we have—

The three sides of any triangle and three concurrent lines through the vertices cut any transversal in a six-point involution.

Exercises.

1. A circle and an inscribed quadrangle cut any line through them in involution.
2. The circles of a co-axal system cut any line through them in involution.
3. Any three concurrent chords intersect the circle in six points forming a system in involution.
4. The circles of a co-axal system cut any other circle in involution.
5. Any four circles through a common point have their six radical axes forming a pencil in involution.

INDEX OF DEFINITIONS, TERMS, ETC.

The Numbers refer to the Articles.

Addition Theorem for Sine and Cosine, . . .	236
Altitude, ...	87
Ambiguous Case, . .	66
Angle, ...	31
,, Acute, . . .	40
,, Adjacent internal,	49
,, Basal, . . .	49
,, External, . .	49
,, Obtuse, . .	40
,, Re-entrant, . .	89
,, Right, . . .	36
,, Straight, . .	36
,, Arms of, . .	32
,, Bisectors of, . .	43
,, Complement of, .	40
,, Cosine of, . .	213
,, Measure of, . 41,	207
,, Sine of, . .	213
,, Supplement of, .	40
,, Tangent of, . .	213
,, Vertex of, . .	32
Angles, Adjacent, . .	35
,, Alternate, . .	73
,, Interadjacent, .	73
,, Opposite, . .	39
Angles, Vertical, . .	39
,, Sum and Difference of,	35
,, of the same Affection,	65
Anharmonic Ratio, . .	298
Antihomologous, . .	289
Apothem, . . .	146
Arc, ...	101
Area, ...	136
,, of a Triangle, .	175½
Axiom, ...	3
Axis of a Range, . .	230
,, Perspective, .	254
,, Similitude, .	294
,, Symmetry, .	101
Basal Angles, . . .	49
Biliteral Notation, . .	22
Bisectors of an Angle, .	43
Brianchon's Theorem, .	320
Brianchon Point, . .	320
Centre of a Circle, . .	92
,, Inversion, .	256
,, Mean Position,	238
,, Perspective, .	254
,, Similitude,	281

Centre-line,	. . .	95	Diameter, . . .	95
Centre-locus,	. . .	129	Difference of Segments, .	29
Centroid,	. . .	85	Dimension, . . .	27
Circle,	. . .	92	Double Point, . .	109, 328
Circle of Antisimilitude,	.	290		
,, Inversion,	.	256	Eidograph, . .	. 211½
,, Similitude,	.	288	End-points, . . .	22
Circumangle,	. .	36	Envelope, . . .	223
Circumcentre,	.	86, 97	Equal, . . .	27, 136
Circumcircle,	. .	97	Equilateral Triangle,	. 53
Circumference,	. .	92	Excircle, ...	131
Circumradius,	. .	97	External Angle, . .	49
Circumscribed Figure,	.	97	Extreme and Mean Ratio,	183
Chord,	. . .	95	Extremes, . . .	193
Chord of Contact,	. .	114		
Co-axal Circles,	. .	273	Finite Line, . . .	21
Collinear Points,	.	131, 247	Finite Point, . . .	21
Commensurable,	. .	150		
Complement of an Angle,		40	Generating Point, . .	69
Concentric Circles,	. .	93	Geometric Mean, .	. 169
Conclusion,	. . .	4	Given Point and Line, .	20
Concurrent Lines,	.	85, 247		
Congruent,	. . .	51	Harmonic Division,	. 208
Concyclic,	. . .	97	Harmonic Ratio, .	. 299
Conjugate Points, etc.,	.	267	Harmonic Systems,	313, 314
Contact of Circles,	. .	291	Homogeneity, . .	. 160
Continuity, Principle of,	.	104	Homographic Systems,	. 327
Corollary,	. . .	8	Homologous Sides, .	. 196
C.-P. Circles,	. . .	274	Homologous Lines and	
Cosine of an Angle,	.	213	Points, . . .	. 196
Constructive Geometry,	.	117	Hypothenuse, . .	88, 168
Curve,	. . .	15	Hypothesis, . . .	4
Datum Line,	. . .	20	Incircle of a Triangle,	. 131
Degree,	. . .	41	Incommensurable, .	. 150
Desargue's Theorem,	.	253	Infinity, . . .	21, 220
Diagonal,	. .	80, 247	Initial Line, . . .	20
Diagonal Scale,	. .	211½	Inscribed Figure, .	. 97

INDEX OF DEFINITIONS, TERMS, ETC.

Interadjacent Angles,	73
Inverse Points,	179, 256
Inverse Figures,	260
Involution,	332
Isosceles Triangle,	53
Join,	167
Limit,	148
Limiting Points,	274
Line,	12
Line in Opposite Senses,	156
Line-segment,	21
Locus,	69
L.-P. Circles,	274
Magnitude,	190
Major and Minor,	102
Maximum,	175
Mean Centre,	238
Mean Proportional,	169
Means,	193
Measure,	150
Median,	55
Median Section,	183
Metrical Geometry,	150
Minimum,	175
Nine-points Circle,	{ 116, Ex. 6 265, Ex. 2 }
Normal Quadrangle,	89
Obtuse Angle,	40
Opposite Angles,	39
Opposite Internal Angles,	49
Origin,	20
Orthocentre,	88
Orthogonally,	115
Orthogonal Projection,	167, 229
Pantagraph,	$211\tfrac{1}{2}$
Parallel Lines,	70
Parallelogram,	80
Pascal's Hexagram,	319
Pascal Line,	319
Peaucellier's Cell,	211
Pencil,	203
Perigon,	36
Perimeter,	146
Perspective,	254
Perspective, Axis of,	254
,, Centre of,	254
Perpendicular,	40
Physical Line,	12
Plane,	10, 17
Plane Geometric Figure,	10
Plane Geometry,	11
Point,	13
Point of Bisection,	30
,, Contact,	109
Point, Double,	109, 328
Pole,	32, 266
Polar,	266
Polar Reciprocal,	268
Polar Circle, Centre, etc.,	266
Polygon,	132
Prime Vector,	20
Projection,	167
Proportion,	192
Proportional Compasses,	$211\tfrac{1}{2}$
Protractor,	123
Quadrangle, Quadrilateral,	80, 89
Quadrilateral, complete,	247

SYNTHETIC GEOMETRY.

Radian,	207
Radical Axis,	178, 273
Radical Centre,	276
Radius,	92
Radius Vector,	32
Range,	230
Ratio,	188
Reciprocation,	321
Reciprocal Segments,	184
Rectangle,	82
Rectilinear Figure,	14
Reductio ad absurdum,	54
Re-entrant Angle,	89
Rhombus,	82
Right Angle,	36
Right Bisector,	42
Rotation of a Line,	32, 222
Rule,	16
Rule of Identity,	7
Scalene Triangle,	57
Secant,	95
Sector,	211½
Segment,	21
Semicircle,	102
Self-conjugate, Self-reciprocal,	268
Sense of a Line,	156
Sense of a Rectangle,	161
Similar Figures,	206
Similar Triangles,	77, 196
Sine of an Angle,	213
Spatial Figure,	19
Square,	82, 121
Straight Angle,	36
,, Line,	14
,, Edge,	16
Sum of Line-segments,	28
Sum of Angles,	35
Superposition,	26
Supplement of an Angle,	40
Surface,	9
Tangent,	109
Tangent of an Angle,	213
Tensor,	189
Tetragram,	247
Theorem,	2
Trammel,	107
Transversal,	73
Trapezoid,	84
Triangle,	48
,, Acute-angled, etc.,	77
,, Base of,	49
,, Equilateral,	53
,, Isosceles,	53
,, Obtuse-angled,	77
,, Scalene,	57
,, Right-Angled,	77
Uniliteral Notation,	22
Unit-area, Unit-length,	150
Versed Sine of an Arc,	176
Vertex of an Angle,	32
Vertex of a Triangle,	49
Vertical Angles,	39

A Catalogue

OF WORKS IN

Mathematics, Science, Geography

AND

History,

PUBLISHED BY

Macmillan & Co.,

BEDFORD STREET, STRAND, LONDON.

15.11.18.

CONTENTS.

MATHEMATICS—   PAGE
- Arithmetic and Mensuration 3
- Algebra 5
- Euclid, and Elementary Geometry 6
- Trigonometry 8
- Higher Mathematics 9

SCIENCE—
- Natural Philosophy 16
- Astronomy 22
- Chemistry 22
- Biology 25
- Medicine 29
- Anthropology 30
- Physical Geography and Geology 30
- Agriculture 31
- Political Economy 32
- Mental and Moral Philosophy 33

GEOGRAPHY 34
- Macmillan's Geographical Series 35

HISTORY 36

29 AND 30, BEDFORD STREET, COVENT GARDEN,
LONDON, W.C., *December*, 1888.

MATHEMATICS.

(1) Arithmetic and Mensuration, (2) Algebra, (3) Euclid and Elementary Geometry, (4) Trigonometry, (5) Higher Mathematics.

ARITHMETIC AND MENSURATION.

Aldis.—THE GREAT GIANT ARITHMOS. A most Elementary Arithmetic for Children. By MARY STEADMAN ALDIS. With Illustrations. Globe 8vo. 2s. 6d.

Army Preliminary Examination, Specimens of Papers Set at the, 1882—87.—With Answers to the Mathematical Questions. Subjects: Arithmetic, Algebra, Euclid, Geometrical Drawing, Geography, French, English Dictation. Crown 8vo. 3s. 6d.

This selection of papers at recent Preliminary Examinations for the admission to the Army are intended for the use of Candidates preparing for these Examinations.

Bradshaw.—EASY EXERCISES IN ARITHMETIC. By GERALD BRADSHAW, M.A., Assistant Master in Clifton College. Globe 8vo. [*In the press.*

Brook-Smith (J.).—ARITHMETIC IN THEORY AND PRACTICE. By J. BROOK-SMITH, M.A., LL.B., St. John's College, Cambridge; Barrister-at-Law; one of the Masters of Cheltenham College. New Edition, revised. Crown 8vo. 4s. 6d.

Candler.—HELP TO ARITHMETIC. Designed for the use of Schools. By H. CANDLER, M.A., Mathematical Master of Uppingham School. Second Edition. Extra fcap. 8vo. 2s. 6d.

Dalton.—RULES AND EXAMPLES IN ARITHMETIC. By the Rev. T. DALTON, M.A., Assistant-Master in Eton College. New Edition, with Answers. 18mo. 2s. 6d.

Goyen.—HIGHER ARITHMETIC AND ELEMENTARY MENSURATION. By P. GOYEN, Inspector of Schools, New Zealand. Crown 8vo. 5s.

Hall and Knight.—ARITHMETICAL EXERCISES AND EXAMINATION PAPERS. With an Appendix containing Questions in LOGARITHMS and MENSURATION. By H. S. HALL, M.A., formerly Scholar of Christ's College, Cambridge, Master of the Military and Engineering Side, Clifton College ; and S. R. KNIGHT, B.A., formerly Scholar of Trinity College, Cambridge, late Assistant Master at Marlborough College, Authors of "Elementary Algebra," "Algebraical Exercises and Examination Papers," and "Higher Algebra," &c. Globe 8vo. 2s. 6d.

Lock.—Works by Rev. J. B. LOCK, M.A., Senior Fellow, Assistant Tutor, and Lecturer in Gonville and Caius Colleges, Cambridge, formerly Master at Eton.

ARITHMETIC FOR SCHOOLS. With Answers and 1000 additional Examples for Exercise. Third Edition, revised. Stereotyped. Globe 8vo. 4s. 6d. Or in Two Parts:—Part I. Up to and including Practice, with Answers. Globe 8vo. 2s. Part II. With Answers and 1000 additional Examples for Exercise. Globe 8vo. 3s.

₊ *The complete book can also be obtained* without answers *at the same price, though in different binding. But the edition* with answers *will always be supplied unless the other is specially asked for.*

A KEY TO MR. LOCK'S "ARITHMETIC FOR SCHOOLS." By the Rev. R. G. WATSON, M.A., formerly Head Master of the Dorset County School. Crown 8vo. 10s. 6d.

ARITHMETIC FOR BEGINNERS. A School Class Book of Commercial Arithmetic. Globe 8vo. 2s. 6d.

[*A Key is in preparation.*

Pedley.—EXERCISES IN ARITHMETIC for the Use of Schools. Containing more than 7,000 original Examples. By S. PEDLEY, late of Tamworth Grammar School. Crown 8vo. 5s. Also in Two Parts 2s. 6d. each.

Smith.—Works by the Rev. BARNARD SMITH, M.A., late Rector of Glaston, Rutland, and Fellow and Senior Bursar of S. Peter's College, Cambridge.

ARITHMETIC AND ALGEBRA, in their Principles and Application ; with numerous systematically arranged Examples taken from the Cambridge Examination Papers, with especial reference to the Ordinary Examination for the B.A. Degree. New Edition, carefully Revised. Crown 8vo. 10s. 6d.

ARITHMETIC FOR SCHOOLS. New Edition. Crown 8vo. 4s. 6d.

A KEY TO THE ARITHMETIC FOR SCHOOLS. New Edition. Crown 8vo. 8s. 6d.

EXERCISES IN ARITHMETIC. Crown 8vo, limp cloth, 2s. With Answers, 2s. 6d. Answers separately, 6d.

SCHOOL CLASS-BOOK OF ARITHMETIC. 18mo, cloth. 3s. Or sold separately, in Three Parts, 1s. each.

KEYS TO SCHOOL CLASS-BOOK OF ARITHMETIC. Parts I., II., and III., 2s. 6d. each.

Smith.—SHILLING BOOK OF ARITHMETIC FOR NATIONAL AND ELEMENTARY SCHOOLS. 18mo, cloth. Or separately, Part I. 2*d*. ; Part II. 3*d*. ; Part III. 7*d*. Answers, 6*d*.
THE SAME, with Answers complete. 18mo, cloth. 1*s*. 6*d*.
KEY TO SHILLING BOOK OF ARITHMETIC. 18mo. 4*s*. 6*d*.
EXAMINATION PAPERS IN ARITHMETIC. 18mo. 1*s*. 6*d*.
The same, with Answers, 18mo, 2*s*. Answers, 6*d*.
KEY TO EXAMINATION PAPERS IN ARITHMETIC. 18mo. 4*s*. 6*d*.
THE METRIC SYSTEM OF ARITHMETIC, ITS PRINCIPLES AND APPLICATIONS, with numerous Examples, written expressly for Standard V. in National Schools. New Edition. 18mo, cloth, sewed. 3*d*.
A CHART OF THE METRIC SYSTEM, on a Sheet, size 42 in. by 34 in. on Roller, mounted and varnished. New Edition. Price 3*s*. 6*d*.
Also a Small Chart on a Card, price 1*d*.
EASY LESSONS IN ARITHMETIC, combining Exercises in Reading, Writing, Spelling, and Dictation. Part I. for Standard I. in National Schools. Crown 8vo. 9*d*.
EXAMINATION CARDS IN ARITHMETIC. (Dedicated to Lord Sandon.) With Answers and Hints.
Standards I. and II. in box, 1*s*. Standards III., IV., and V., in boxes, 1*s*. each. Standard VI. in Two Parts, in boxes, 1*s*. each.
A and B papers, of nearly the same difficulty, are given so as to prevent copying, and the colours of the A and B papers differ in each Standard, and from those of every other Standard, so that a master or mistress can see at a glance whether the children have the proper papers.

Todhunter.—MENSURATION FOR BEGINNERS. By I. TODHUNTER, M.A., F.R.S., D.Sc., late of St. John's College, Cambridge. With Examples. New Edition. 18mo. 2*s*. 6*d*.
KEY TO MENSURATION FOR BEGINNERS. By the Rev. FR. LAWRENCE MCCARTHY, Professor of Mathematics in St. Peter's College, Agra. Crown 8vo. 7*s*. 6*d*.

ALGEBRA.

Dalton.—RULES AND EXAMPLES IN ALGEBRA. By the Rev. T. DALTON, M.A., Assistant-Master of Eton College. Part I. New Edition. 18mo. 2*s*. Part II. 18mo. 2*s*. 6*d*.
* *A Key to Part I. for Teachers only*, 7*s*. 6*d*.

Hall and Knight.—ELEMENTARY ALGEBRA FOR SCHOOLS. By H. S. HALL, M.A., formerly Scholar of Christ's College, Cambridge, Master of the Military and Engineering Side, Clifton College ; and S. R. KNIGHT, B.A., formerly Scholar of Trinity College, Cambridge, late Assistant-Master at Marlborough College. Fourth Edition, Revised and Corrected. Globe 8vo, bound in maroon coloured cloth, 3*s*. 6*d*. ; with Answers, bound in green coloured cloth, 4*s*. 6*d*.

Hall and Knight.—ALGEBRAICAL EXERCISES AND EXAMINATION PAPERS. To accompany ELEMENTARY ALGEBRA. Second Edition, revised. Globe 8vo. 2s. 6d.
HIGHER ALGEBRA. A Sequel to "ELEMENTARY ALGEBRA FOR SCHOOLS." Second Edition. Crown 8vo. 7s. 6d.

Jones and Cheyne.—ALGEBRAICAL EXERCISES. Progressively Arranged. By the Rev. C. A. JONES, M.A., and C. H. CHEYNE, M.A., F.R.A.S., Mathematical Masters of Westminster School. New Edition. 18mo. 2s. 6d.
SOLUTIONS AND HINTS FOR THE SOLUTION OF SOME OF THE EXAMPLES IN THE ALGEBRAICAL EXERCISES OF MESSRS. JONES AND CHEYNE. By Rev. W. FAILES, M.A., Mathematical Master at Westminster School, late Scholar of Trinity College, Cambridge. Crown 8vo. 7s. 6d.

Smith (Barnard).—ARITHMETIC AND ALGEBRA, in their Principles and Application; with numerous systematically arranged Examples taken from the Cambridge Examination Papers, with especial reference to the Ordinary Examination for the B.A. Degree. By the Rev. BARNARD SMITH, M.A., late Rector of Glaston, Rutland, and Fellow and Senior Bursar of St. Peter's College, Cambridge. New Edition, carefully Revised. Crown 8vo. 10s. 6d.

Smith (Charles).—Works by CHARLES SMITH, M.A., Fellow and Tutor of Sidney Sussex College, Cambridge.
ELEMENTARY ALGEBRA. Globe 8vo. 4s. 6d.
In this work the author has endeavoured to explain the principles of Algebra in as simple a manner as possible for the benefit of beginners, bestowing great care upon the explanations and proofs of the fundamental operations and rules.
A TREATISE ON ALGEBRA. Crown 8vo. 7s. 6d.
[*A Key is in the press, nearly ready.*

Todhunter.—Works by I. TODHUNTER, M.A., F.R.S., D.Sc., late of St. John's College, Cambridge.
"Mr. Todhunter is chiefly known to Students of Mathematics as the author of a series of admirable mathematical text-books, which possess the rare qualities of being clear in style and absolutely free from mistakes, typographical or other."—SATURDAY REVIEW.
ALGEBRA FOR BEGINNERS. With numerous Examples. New Edition. 18mo. 2s. 6d.
KEY TO ALGEBRA FOR BEGINNERS. Crown 8vo. 6s. 6d.
ALGEBRA. For the Use of Colleges and Schools. New Edition. Crown 8vo. 7s. 6d.
KEY TO ALGEBRA FOR THE USE OF COLLEGES AND SCHOOLS. Crown 8vo. 10s. 6d.

EUCLID, & ELEMENTARY GEOMETRY.

Constable.—GEOMETRICAL EXERCISES FOR BEGINNERS. By SAMUEL CONSTABLE. Crown 8vo. 3s. 6d.
Cuthbertson.—EUCLIDIAN GEOMETRY. By FRANCIS CUTHBERTSON, M.A., LL.D., Head Mathematical Master of the City of London School. Extra fcap. 8vo. 4s. 6d.

Dodgson.—Works by CHARLES L. DODGSON, M.A., Student and late Mathematical Lecturer of Christ Church, Oxford.

    EUCLID. BOOKS I. AND II. Fourth Edition, with words substituted for the Algebraical Symbols used in the First Edition. Crown 8vo. 2s.

    ⁎ The text of this Edition has been ascertained, by counting the words, to be *less than five-sevenths* of that contained in the ordinary editions.

    EUCLID AND HIS MODERN RIVALS. Second Edition. Crown 8vo. 6s.

    CURIOSA MATHEMATICA. Part I. A New Theory of Parallels. Crown 8vo. 2s.

Dupuis.—ELEMENTARY SYNTHETIC GEOMETRY OF THE POINT, LINE, AND CIRCLE IN THE PLANE. By N. F. DUPUIS, M.A., F.R.S.C., Professor of Pure Mathematics in the University of Queen's College, Kingston, Canada. Globe 8vo. [*In the press.*

Eagles.—CONSTRUCTIVE GEOMETRY OF PLANE CURVES. By T. H. EAGLES, M.A., Instructor in Geometrical Drawing, and Lecturer in Architecture at the Royal Indian Engineering College, Cooper's Hill. With numerous Examples. Crown 8vo. 12s.

Hall and Stevens.—A TEXT BOOK OF EUCLID'S ELEMENTS. Including alternative Proofs, together with additional Theorems and Exercises, classified and arranged. By H. S. HALL, M.A., formerly Scholar of Christ's College, Cambridge, and F. H. STEVENS, M.A., formerly Scholar of Queen's College, Oxford: Masters of the Military and Engineering Side, Clifton College. Globe 8vo. Books I.—VI. 4s. 6d. Or in Two Parts. Part I., Books I. and II. Second Edition. 2s. Part II., Books III.—VI. 3s.

Halsted.—THE ELEMENTS OF GEOMETRY. By GEORGE BRUCE HALSTED, Professor of Pure and Applied Mathematics in the University of Texas. 8vo. 12s. 6d.

Kitchener.—A GEOMETRICAL NOTE-BOOK, containing Easy Problems in Geometrical Drawing preparatory to the Study of Geometry. For the Use of Schools. By F. E. KITCHENER, M.A., Head-Master of the Grammar School, Newcastle, Staffordshire. New Edition. 4to. 2s.

Lock.—EUCLID FOR BEGINNERS. Being an Introduction to existing Text-books. By Rev. J. B. LOCK, M.A., author of "Arithmetic for Schools," "Trigonometry," "Dynamics," "Elementary Statics," &c. [*In preparation.*

Mault.—NATURAL GEOMETRY: an Introduction to the Logical Study of Mathematics. For Schools and Technical Classes. With Explanatory Models, based upon the Tachymetrical works of Ed. Lagout. By A. MAULT. 18mo. 1s.

    Models to Illustrate the above, in Box, 12s. 6d.

Millar.—ELEMENTS OF DESCRIPTIVE GEOMETRY. By J. B. MILLAR, M.E., Civil Engineer, Lecturer on Engineering in the Victoria University, Manchester. Second Edition. Cr. 8vo. 6s.

Syllabus of Plane Geometry (corresponding to Euclid, Books I.—VI.). Prepared by the Association for the Improvement of Geometrical Teaching. New Edition. Crown 8vo. 1s.

Todhunter.—THE ELEMENTS OF EUCLID. For the Use of Colleges and Schools. By I. TODHUNTER, M.A., F.R.S., D.Sc., of St. John's College, Cambridge. New Edition. 18mo. 3s 6d.

KEY TO EXERCISES IN EUCLID. Crown 8vo. 6s. 6d.

Wilson (J. M.).—ELEMENTARY GEOMETRY. BOOKS I.—V. Containing the Subjects of Euclid's first Six Books. Following the Syllabus of the Geometrical Association. By the Rev. J. M. WILSON, M.A., Head Master of Clifton College. New Edition. Extra fcap. 8vo. 4s. 6d.

TRIGONOMETRY.

Beasley.—AN ELEMENTARY TREATISE ON PLANE TRIGONOMETRY. With Examples. By R. D. BEASLEY, M.A. Ninth Edition, revised and enlarged. Crown 8vo. 3s. 6d.

Johnson.—A TREATISE ON TRIGONOMETRY. By W. E. JOHNSON, M.A., formerly Scholar of King's College, Cambridge. Crown 8vo. [*In the press.*

Lock.—Works by Rev. J. B. LOCK, M.A., Senior Fellow, Assistant Tutor and Lecturer in Gonville and Caius Colleges, Cambridge; formerly Master at Eton.

TRIGONOMETRY FOR BEGINNERS, as far as the Solution of Triangles. Globe 8vo. 2s. 6d.

ELEMENTARY TRIGONOMETRY. Sixth Edition (in this edition the chapter on logarithms has been carefully revised.) Globe 8vo. 4s. 6d. [*A Key is in the press.*

Mr. E. J. ROUTH, D.Sc., F.R.S., writes:—"It is an able treatise. It takes the difficulties of the subject one at a time, and so leads the young student easily along."

HIGHER TRIGONOMETRY. Fifth Edition. Globe 8vo. 4s. 6d. Both Parts complete in One Volume. Globe 8vo. 7s. 6d.

(See also under *Arithmetic*, *Higher Mathematics*, and *Euclid*.)

M'Clelland and Preston.—A TREATISE ON SPHERICAL TRIGONOMETRY. With numerous Examples. By WILLIAM J. M'CLELLAND, Sch.B.A., Principal of the Incorporated Society's School, Santry, Dublin, and THOMAS PRESTON, Sch.B.A. In Two Parts. Crown 8vo. Part I. To the End of Solution of Triangles, 4s. 6d. Part II., 5s.

Palmer.—TEXT-BOOK OF PRACTICAL LOGARITHMS AND TRIGONOMETRY. By J. H. PALMER, Head Schoolmaster R.N., H.M.S. *Cambridge*, Devonport. Globe 8vo. 4s. 6d.

MATHEMATICS.

Snowball.—THE ELEMENTS OF PLANE AND SPHERICAL TRIGONOMETRY. By J. C. SNOWBALL, M.A. Fourteenth Edition. Crown 8vo. 7s. 6d.

Todhunter.—Works by I. TODHUNTER, M.A., F.R.S., D.Sc., late of St. John's College, Cambridge.
  TRIGONOMETRY FOR BEGINNERS. With numerous Examples. New Edition. 18mo. 2s. 6d.
  KEY TO TRIGONOMETRY FOR BEGINNERS. Cr. 8vo. 8s. 6d.
  PLANE TRIGONOMETRY. For Schools and Colleges. New Edition. Crown 8vo. 5s.
  KEY TO PLANE TRIGONOMETRY. Crown 8vo. 10s. 6d.
  A TREATISE ON SPHERICAL TRIGONOMETRY. New Edition, enlarged. Crown 8vo. 4s. 6d.
  (See also under *Arithmetic and Mensuration*, *Algebra*, and *Higher Mathematics*.)

HIGHER MATHEMATICS.

Airy.—Works by Sir G. B. AIRY, K.C.B., formerly Astronomer-Royal.
  ELEMENTARY TREATISE ON PARTIAL DIFFERENTIAL EQUATIONS. Designed for the Use of Students in the Universities. With Diagrams. Second Edition. Crown 8vo. 5s. 6d.
  ON THE ALGEBRAICAL AND NUMERICAL THEORY OF ERRORS OF OBSERVATIONS AND THE COMBINATION OF OBSERVATIONS. Second Edition, revised. Crown 8vo. 6s. 6d.

Alexander (T.).—ELEMENTARY APPLIED MECHANICS. Being the simpler and more practical Cases of Stress and Strain wrought out individually from first principles by means of Elementary Mathematics. By T. ALEXANDER, C.E., Professor of Civil Engineering in the Imperial College of Engineering, Tokei, Japan. Part I. Crown 8vo. 4s. 6d.

Alexander and Thomson.—ELEMENTARY APPLIED MECHANICS. By THOMAS ALEXANDER, C.E., Professor of Engineering in the Imperial College of Engineering, Tokei, Japan; and ARTHUR WATSON THOMSON, C.E., B.Sc., Professor of Engineering at the Royal College, Cirencester. Part II. TRANSVERSE STRESS. Crown 8vo. 10s. 6d.

Army Preliminary Examination, 1882-1887, Specimens of Papers set at the. With answers to the Mathematical Questions. Subjects: Arithmetic, Algebra, Euclid, Geometrical Drawing, Geography, French, English Dictation. Cr. 8vo. 3s. 6d.

Ball (W. W. R.).—A SHORT ACCOUNT OF THE HISTORY OF MATHEMATICS. By W. W. ROUSE BALL, Fellow and Assistant Tutor of Trinity College, Cambridge, and of the Inner Temple, Barrister-at-Law. Crown 8vo. 10s. 6d.

Ball (Sir R. S.).—EXPERIMENTAL MECHANICS. A Course of Lectures delivered at the Royal College of Science for Ireland. By Sir Robert Stawell Ball, LL.D., F.R.S., Astronomer-Royal of Ireland. New Edition. With Illustrations. Crown 8vo. 6s.

Boole.—THE CALCULUS OF FINITE DIFFERENCES. By G. Boole, D.C.L., F.R.S., late Professor of Mathematics in the Queen's University, Ireland. Third Edition, revised by J. F. Moulton. Crown 8vo. 10s. 6d.

Cambridge Senate-House Problems and Riders, with Solutions:—

1875—PROBLEMS AND RIDERS. By A. G. Greenhill, M.A. Crown 8vo. 8s. 6d.

1878—SOLUTIONS OF SENATE-HOUSE PROBLEMS. By the Mathematical Moderators and Examiners. Edited by J. W. L. Glaisher, M.A., Fellow of Trinity College, Cambridge. 12s.

Carll.—A TREATISE ON THE CALCULUS OF VARIATIONS. Arranged with the purpose of Introducing, as well as Illustrating, its Principles to the Reader by means of Problems, and Designed to present in all Important Particulars a Complete View of the Present State of the Science. By Lewis Buffett Carll, A.M. Demy 8vo. 21s.

Cheyne.—AN ELEMENTARY TREATISE ON THE PLANETARY THEORY. By C. H. H. Cheyne, M.A., F.R.A.S. With a Collection of Problems. Third Edition. Edited by Rev. A. Freeman, M.A., F.R.A.S. Crown 8vo. 7s. 6d.

Christie.—A COLLECTION OF ELEMENTARY TEST-QUESTIONS IN PURE AND MIXED MATHEMATICS; with Answers and Appendices on Synthetic Division, and on the Solution of Numerical Equations by Horner's Method. By James R. Christie, F.R.S., Royal Military Academy, Woolwich. Crown 8vo. 8s. 6d.

Clausius.—MECHANICAL THEORY OF HEAT. By R. Clausius. Translated by Walter R. Browne, M.A., late Fellow of Trinity College, Cambridge. Crown 8vo. 10s. 6d.

Clifford.—THE ELEMENTS OF DYNAMIC. An Introduction to the Study of Motion and Rest in Solid and Fluid Bodies. By W. K. Clifford, F.R.S., late Professor of Applied Mathematics and Mechanics at University College, London. Part I.—KINEMATIC. Crown 8vo. Books I—III. 7s. 6d.; Book IV. and Appendix 6s.

Cockshott and Walters.—GEOMETRICAL CONICS. An Elementary Treatise. Drawn up in accordance with the Syllabus issued by the Society for the Improvement of Geometrical Teaching. By A. Cockshott, M.A., formerly Fellow and Assistant-Tutor of Trinity College, Cambridge, and Assistant-Master at Eton; and Rev. F. B. Walters, M.A., Fellow of Queens' College, Cambridge, and Principal of King William's College, Isle of Man. With Diagrams. Crown 8vo.
[*In the press.*

MATHEMATICS.

Cotterill.—APPLIED MECHANICS: an Elementary General Introduction to the Theory of Structures and Machines. By JAMES H. COTTERILL, F.R.S., Associate Member of the Council of the Institution of Naval Architects, Associate Member of the Institution of Civil Engineers, Professor of Applied Mechanics in the Royal Naval College, Greenwich. Medium 8vo. 18s.

ELEMENTARY MANUAL OF APPLIED MECHANICS. By the same Author. Crown 8vo. [*In preparation.*

Day (R. E.)—ELECTRIC LIGHT ARITHMETIC. By R. E. DAY, M.A., Evening Lecturer in Experimental Physics at King's College, London. Pott 8vo. 2s.

Dodgson.—CURIOSA MATHEMATICA. Part I. A New Theory of Parallels. By CHARLES L. DODGSON, M.A., Student and late Mathematical Lecturer of Christ Church, Oxford, Author of "Euclid and his Modern Rivals," "Euclid, Books I. and II.," &c. Crown 8vo. 2s.

Drew.—GEOMETRICAL TREATISE ON CONIC SECTIONS. By W. H. DREW, M.A., St. John's College, Cambridge. New Edition, enlarged. Crown 8vo. 5s.

Dyer.—EXERCISES IN ANALYTICAL GEOMETRY. Compiled and arranged by J. M. DYER, M.A. With Illustrations. Crown 8vo. 4s. 6d.

Eagles.—CONSTRUCTIVE GEOMETRY OF PLANE CURVES. By T. H. EAGLES, M.A., Instructor in Geometrical Drawing, and Lecturer in Architecture at the Royal Indian Engineering College, Cooper's Hill. With numerous Examples. Crown 8vo. 12s.

Edgar (J. H.) and Pritchard (G. S.).—NOTE-BOOK ON PRACTICAL SOLID OR DESCRIPTIVE GEOMETRY. Containing Problems with help for Solutions. By J. H. EDGAR, M.A., Lecturer on Mechanical Drawing at the Royal School of Mines, and G. S. PRITCHARD. Fourth Edition, revised by ARTHUR MEEZE. Globe 8vo. 4s. 6d.

Edwards.—THE DIFFERENTIAL CALCULUS. With Applications and numerous Examples. An Elementary Treatise by JOSEPH EDWARDS, M.A., formerly Fellow of Sidney Sussex College, Cambridge. Crown 8vo. 10s. 6d.

Ferrers.—Works by the Rev. N. M. FERRERS, M.A., Master of Gonville and Caius College, Cambridge.

AN ELEMENTARY TREATISE ON TRILINEAR CO-ORDINATES, the Method of Reciprocal Polars, and the Theory of Projectors. New Edition, revised. Crown 8vo. 6s. 6d.

AN ELEMENTARY TREATISE ON SPHERICAL HARMONICS, AND SUBJECTS CONNECTED WITH THEM. Crown 8vo. 7s. 6d.

Forsyth.—A TREATISE ON DIFFERENTIAL EQUATIONS. By ANDREW RUSSELL FORSYTH, M.A., F.R.S., Fellow and Assistant Tutor of Trinity College, Cambridge. 8vo. 14s.

Frost.—Works by PERCIVAL FROST, M.A., D.Sc., formerly Fellow of St. John's College, Cambridge; Mathematical Lecturer at King's College.
  AN ELEMENTARY TREATISE ON CURVE TRACING. 8vo. 12s.
  SOLID GEOMETRY. Third Edition. Demy 8vo. 16s.
  HINTS FOR THE SOLUTION OF PROBLEMS in the Third Edition of SOLID GEOMETRY. 8vo. 8s. 6d.

Greaves.—A TREATISE ON ELEMENTARY STATICS. By JOHN GREAVES, M.A., Fellow and Mathematical Lecturer of Christ's College, Cambridge. Second Edition. Crown 8vo. 6s. 6d.
  STATICS FOR SCHOOLS. By the Same Author. [*In the press.*

Greenhill.—DIFFERENTIAL AND INTEGRAL CALCULUS. With Applications. By A. G. GREENHILL, M.A., Professor of Mathematics to the Senior Class of Artillery Officers, Woolwich, and Examiner in Mathematics to the University of London. Crown 8vo. 7s. 6d.

Hemming.—AN ELEMENTARY TREATISE ON THE DIFFERENTIAL AND INTEGRAL CALCULUS, for the Use of Colleges and Schools. By G. W. HEMMING, M.A., Fellow of St. John's College, Cambridge. Second Edition, with Corrections and Additions. 8vo. 9s.

Ibbetson.—THE MATHEMATICAL THEORY OF PERFECTLY ELASTIC SOLIDS, with a short account of Viscous Fluids. An Elementary Treatise. By WILLIAM JOHN IBBETSON, M.A., Fellow of the Royal Astronomical Society, and of the Cambridge Philosophical Society, Member of the London Mathematical Society, late Senior Scholar of Clare College, Cambridge. 8vo. 21s.

Jellett (John H.).—A TREATISE ON THE THEORY OF FRICTION. By JOHN H. JELLETT, B.D., late Provost of Trinity College, Dublin; President of the Royal Irish Academy. 8vo. 8s. 6d.

Johnson.—Works by WILLIAM WOOLSEY JOHNSON, Professor of Mathematics at the U.S. Naval Academy, Annopolis, Maryland.
  INTEGRAL CALCULUS, an Elementary Treatise on the; Founded on the Method of Rates or Fluxions. Demy 8vo. 9s.
  CURVE TRACING IN CARTESIAN CO-ORDINATES. Crown 8vo. 4s. 6d.

Jones.—EXAMPLES IN PHYSICS. By D. E. JONES, B.Sc., Lecturer in Physics in University College of Wales, Aberystwyth. Fcap. 8vo. 3s. 6d.

Kelland and Tait.—INTRODUCTION TO QUATERNIONS, with numerous examples. By P. KELLAND, M.A., F.R.S., and P. G. TAIT, M.A., Professors in the Department of Mathematics in the University of Edinburgh. Second Edition. Crown 8vo. 7s. 6d.

Kempe.—HOW TO DRAW A STRAIGHT LINE: a Lecture on Linkages. By A. B. KEMPE. With Illustrations. Crown 8vo. 1s. 6d. (*Nature Series.*)

MATHEMATICS.

Kennedy.—THE MECHANICS OF MACHINERY. By A. B. W. KENNEDY, F.R.S., M.Inst.C.E., Professor of Engineering and Mechanical Technology in University College, London. With Illustrations. Crown 8vo. 12s. 6d.

Knox.—DIFFERENTIAL CALCULUS FOR BEGINNERS. By ALEXANDER KNOX. Fcap. 8vo. 3s. 6d.

Lock.—Works by the Rev. J. B. LOCK, M.A., Author of "Trigonometry," "Arithmetic for Schools," &c.
HIGHER TRIGONOMETRY. Sixth Edition. Globe 8vo. 4s. 6d.
DYNAMICS FOR BEGINNERS. Second Edition. (Stereotyped.) Globe 8vo. 3s. 6d.
ELEMENTARY STATICS. Globe 8vo.
(See also under *Arithmetic*, *Euclid*, and *Trigonometry*.)

Lupton.—CHEMICAL ARITHMETIC. With 1,200 Examples. By SYDNEY LUPTON, M.A., F.C.S., F.I.C., formerly Assistant-Master in Harrow School. Second Edition. Fcap. 8vo. 4s. 6d.

Macfarlane.—PHYSICAL ARITHMETIC. By ALEXANDER MACFARLANE, M.A., D.Sc., F.R.S.E., Examiner in Mathematics to the University of Edinburgh. Crown 8vo. 7s. 6d.

MacGregor.—KINEMATICS AND DYNAMICS. An Elementary Treatise. By JAMES GORDON MACGREGOR, M.A., D.Sc., Fellow of the Royal Societies of Edinburgh and of Canada, Munro Professor of Physics in Dalhousie College, Halifax, Nova Scotia. With Illustrations. Crown 8vo. 10s. 6d.

Merriman.—A TEXT BOOK OF THE METHOD OF LEAST SQUARES. By MANSFIELD MERRIMAN, Professor of Civil Engineering at Lehigh University, Member of the American Philosophical Society, American Association for the Advancement of Science, &c. Demy 8vo. 8s. 6d.

Millar.—ELEMENTS OF DESCRIPTIVE GEOMETRY. By J. B. MILLAR, C.E., Assistant Lecturer in Engineering in Owens College, Manchester. Second Edition. Crown 8vo. 6s.

Milne.—Works by the Rev. JOHN J. MILNE, M.A., Private Tutor, late Scholar, of St. John's College, Cambridge, &c., &c., formerly Second Master of Heversham Grammar School.
WEEKLY PROBLEM PAPERS. With Notes intended for the use of students preparing for Mathematical Scholarships, and for the Junior Members of the Universities who are reading for Mathematical Honours. Pott 8vo. 4s. 6d.
SOLUTIONS TO WEEKLY PROBLEM PAPERS. Crown 8vo. 10s. 6d.
COMPANION TO "WEEKLY PROBLEM PAPERS." Crown 8vo. 10s. 6d.

Muir.—A TREATISE ON THE THEORY OF DETERMINANTS. With graduated sets of Examples. For use in Colleges and Schools. By THOS. MUIR, M.A., F.R.S.E., Mathematical Master in the High School of Glasgow. Crown 8vo. 7s. 6d.

Palmer.—TEXT BOOK OF PRACTICAL LOGARITHMS AND TRIGONOMETRY. By J. H. PALMER, Head Schoolmaster R.N., H.M.S. *Cambridge*, Devonport. Globe 8vo. 4s. 6d.

Parkinson.—AN ELEMENTARY TREATISE ON MECHANICS. For the Use of the Junior Classes at the University and the Higher Classes in Schools. By S. PARKINSON, D.D., F.R.S., Tutor and Prælector of St. John's College, Cambridge. With a Collection of Examples. Sixth Edition, revised. Crown 8vo. 9s. 6d.

Pirie.—LESSONS ON RIGID DYNAMICS. By the Rev. G. PIRIE, M.A., late Fellow and Tutor of Queen's College, Cambridge; Professor of Mathematics in the University of Aberdeen. Crown 8vo. 6s.

Puckle.—AN ELEMENTARY TREATISE ON CONIC SECTIONS AND ALGEBRAIC GEOMETRY. With Numerous Examples and Hints for their Solution; especially designed for the Use of Beginners. By G. H. PUCKLE, M.A. Fifth Edition, revised and enlarged. Crown 8vo. 7s. 6d.

Reuleaux.—THE KINEMATICS OF MACHINERY. Outlines of a Theory of Machines. By Professor F. REULEAUX Translated and Edited by Professor A. B. W. KENNEDY, F.R.S. C.E. With 450 Illustrations. Medium 8vo. 21s.

Rice and Johnson.—DIFFERENTIAL CALCULUS, an Elementary Treatise on the; Founded on the Method of Rates or Fluxions. By JOHN MINOT RICE, Professor of Mathematics in the United States Navy, and WILLIAM WOOLSEY JOHNSON, Professor of Mathematics at the United States Naval Academy. Third Edition, Revised and Corrected. Demy 8vo. 18s. Abridged Edition, 9s.

Robinson.—TREATISE ON MARINE SURVEYING. Prepared for the use of younger Naval Officers. With Questions for Examinations and Exercises principally from the Papers of the Royal Naval College. With the results. By Rev. JOHN L. ROBINSON, Chaplain and Instructor in the Royal Naval College, Greenwich. With Illustrations. Crown 8vo. 7s. 6d.

CONTENTS.—Symbols used in Charts and Surveying—The Construction and Use of Scales—Laying off Angles—Fixing Positions by Angles — Charts and Chart-Drawing—Instruments and Observing — Base Lines—Triangulation—Levelling—Tides and Tidal Observations—Soundings—Chronometers—Meridian Distances—Method of Plotting a Survey—Miscellaneous Exercises—Index.

Routh.—Works by EDWARD JOHN ROUTH, D.Sc., LL.D., F.R.S., Fellow of the University of London, Hon. Fellow of St. Peter's College, Cambridge.
A TREATISE ON THE DYNAMICS OF THE SYSTEM OF RIGID BODIES. With numerous Examples. Fourth and enlarged Edition. Two Vols. 8vo. Vol. I.—Elementary Parts. 14s. Vol. II.—The Advanced Parts. 14s.
STABILITY OF A GIVEN STATE OF MOTION, PARTICULARLY STEADY MOTION. Adams' Prize Essay for 1877. 8vo. 8s. 6d.

MATHEMATICS.

Smith (C.).—Works by CHARLES SMITH, M.A., Fellow and Tutor of Sidney Sussex College, Cambridge.
CONIC SECTIONS. Fourth Edition. Crown 8vo. 7s. 6d.
SOLUTIONS TO CONIC SECTIONS. Crown 8vo. 10s. 6d.
AN ELEMENTARY TREATISE ON SOLID GEOMETRY. Second Edition. Crown 8vo. 9s. 6d. (See also under *Algebra*.)

Tait and Steele.—A TREATISE ON DYNAMICS OF A PARTICLE. With numerous Examples. By Professor TAIT and Mr. STEELE. Fifth Edition, revised. Crown 8vo. 12s.

Thomson.—Works by J. J. THOMSON, Fellow of Trinity College, Cambridge, and Professor of Experimental Physics in the University.
A TREATISE ON THE MOTION OF VORTEX RINGS. An Essay to which the Adams Prize was adjudged in 1882 in the University of Cambridge. With Diagrams. 8vo. 6s.
APPLICATIONS OF DYNAMICS TO PHYSICS AND CHEMISTRY. Crown 8vo. 7s. 6d.

Todhunter.—Works by I. TODHUNTER, M.A., F.R.S., D.Sc., late of St. John's College, Cambridge.

"Mr. Todhunter is chiefly known to students of Mathematics as the author of a series of admirable mathematical text-books, which possess the rare qualities of being clear in style and absolutely free from mistakes, typographical and other."—SATURDAY REVIEW.

MECHANICS FOR BEGINNERS. With numerous Examples. New Edition. 18mo. 4s. 6d.
KEY TO MECHANICS FOR BEGINNERS. Crown 8vo. 6s. 6d.
AN ELEMENTARY TREATISE ON THE THEORY OF EQUATIONS. New Edition, revised. Crown 8vo. 7s. 6d.
PLANE CO-ORDINATE GEOMETRY, as applied to the Straight Line and the Conic Sections. With numerous Examples. New Edition, revised and enlarged. Crown 8vo. 7s. 6d.
KEY TO PLANE CO-ORDINATE GEOMETRY. By C. W. BOURNE, M.A. Head Master of the College, Inverness. Crown 8vo. 10s. 6d.
A TREATISE ON THE DIFFERENTIAL CALCULUS. With numerous Examples. New Edition. Crown 8vo. 10s. 6d.
A KEY TO DIFFERENTIAL CALCULUS. By H. ST. J. HUNTER, M.A. Crown 8vo. 10s. 6d.
A TREATISE ON THE INTEGRAL CALCULUS AND ITS APPLICATIONS. With numerous Examples. New Edition, revised and enlarged. Crown 8vo. 10s. 6d.
EXAMPLES OF ANALYTICAL GEOMETRY OF THREE DIMENSIONS. New Edition, revised. Crown 8vo. 4s.
A TREATISE ON ANALYTICAL STATICS. With numerous Examples. Fifth Edition. Edited by Professor J. D. EVERETT, F.R.S. Crown 8vo. 10s. 6d.
A HISTORY OF THE MATHEMATICAL THEORY OF PROBABILITY, from the time of Pascal to that of Laplace. 8vo. 18s.

Todhunter.—*continued.*
A HISTORY OF THE MATHEMATICAL THEORIES OF ATTRACTION, AND THE FIGURE OF THE EARTH, from the time of Newton to that of Laplace. 2 vols. 8vo. 24s.

AN ELEMENTARY TREATISE ON LAPLACE'S, LAME'S, AND BESSEL'S FUNCTIONS. Crown 8vo. 10s. 6d.

(See also under *Arithmetic and Mensuration, Algebra,* and *Trigonometry.*)

Wilson (J. M.).—SOLID GEOMETRY AND CONIC SECTIONS. With Appendices on Transversals and Harmonic Division. For the Use of Schools. By Rev. J. M. WILSON, M.A. Head Master of Clifton College. New Edition. Extra fcap. 8vo. 3s. 6d.

Woolwich Mathematical Papers, for Admission into the Royal Military Academy, Woolwich, 1880—1884 inclusive. Crown 8vo. 3s. 6d.

Wolstenholme.—MATHEMATICAL PROBLEMS, on Subjects included in the First and Second Divisions of the Schedule of subjects for the Cambridge Mathematical Tripos Examination. Devised and arranged by JOSEPH WOLSTENHOLME, D.Sc., late Fellow of Christ's College, sometime Fellow of St. John's College, and Professor of Mathematics in the Royal Indian Engineering College. New Edition, greatly enlarged. 8vo. 18s.

EXAMPLES FOR PRACTICE IN THE USE OF SEVEN-FIGURE LOGARITHMS. For the Use of Colleges and Schools. By the same Author. 8vo. 5s.

SCIENCE.

(1) Natural Philosophy, (2) Astronomy, (3) Chemistry, (4) Biology, (5) Medicine, (6) Anthropology, (7) Physical Geography and Geology, (8) Agriculture.

NATURAL PHILOSOPHY.

Airy.—Works by Sir G. B. AIRY, K.C.B., formerly Astronomer-Royal.
ON SOUND AND ATMOSPHERIC VIBRATIONS. With the Mathematical Elements of Music. Designed for the Use of Students in the University. Second Edition, revised and enlarged. Crown 8vo. 9s.

A TREATISE ON MAGNETISM. Designed for the Use of Students in the University. Crown 8vo. 9s. 6d.

GRAVITATION: an Elementary Explanation of the Principal Perturbations in the Solar System. Second Edition. Crown 8vo. 7s. 6d.

SCIENCE.

Alexander (T.).—ELEMENTARY APPLIED MECHANICS. Being the simpler and more practical Cases of Stress and Strain wrought out individually from first principles by means of Elementary Mathematics. By T. ALEXANDER, C.E., Professor of Civil Engineering in the Imperial College of Engineering, Tokei, Japan. Crown 8vo. Part I. 4s. 6d.

Alexander — Thomson.—ELEMENTARY APPLIED MECHANICS. By THOMAS ALEXANDER, C.E., Professor of Engineering in the Imperial College of Engineering, Tokei, Japan; and ARTHUR WATSON THOMSON, C.E., B.Sc., Professor of Engineering at the Royal College, Cirencester. Part II. TRANSVERSE STRESS; upwards of 150 Diagrams, and 200 Examples carefully worked out. Crown 8vo. 10s. 6d.

Ball (R. S.).—EXPERIMENTAL MECHANICS. A Course of Lectures delivered at the Royal College of Science for Ireland. By Sir R. S. BALL, LL.D., F.R.S., Astronomer Royal of Ireland. Second and Cheaper Edition. Crown 8vo. 6s.

Bottomley.—FOUR-FIGURE MATHEMATICAL TABLES. Comprising Logarithmic and Trigonometrical Tables, and Tables of Squares, Square Roots, and Reciprocals. By J. T. BOTTOMLEY, M.A., F.R.S.E., F.C.S., Lecturer in Natural Philosophy in the University of Glasgow. 8vo. 2s. 6d.

Chisholm.—THE SCIENCE OF WEIGHING AND MEASURING, AND THE STANDARDS OF MEASURE AND WEIGHT. By H. W. CHISHOLM, Warden of the Standards. With numerous Illustrations. Crown 8vo. 4s. 6d. (*Nature Series*).

Clausius.—MECHANICAL THEORY OF HEAT. By R. CLAUSIUS. Translated by WALTER R. BROWNE, M.A., late Fellow of Trinity College, Cambridge. Crown 8vo. 10s. 6d.

Cotterill.—APPLIED MECHANICS: an Elementary General Introduction to the Theory of Structures and Machines. By JAMES H. COTTERILL, F.R.S., Associate Member of the Council of the Institution of Naval Architects, Associate Member of the Institution of Civil Engineers, Professor of Applied Mechanics in the Royal Naval College, Greenwich. Medium 8vo. 18s.
ELEMENTARY MANUAL OF APPLIED MECHANICS. By the same Author. Crown 8vo. [*In preparation.*

Cumming.—AN INTRODUCTION TO THE THEORY OF ELECTRICITY. By LINNÆUS CUMMING, M.A., one of the Masters of Rugby School. With Illustrations. Crown 8vo. 8s. 6d.

Daniell.—A TEXT-BOOK OF THE PRINCIPLES OF PHYSICS. By ALFRED DANIELL, M.A., LL.B., D.Sc., F.R.S.E., late Lecturer on Physics in the School of Medicine, Edinburgh. With Illustrations. Second Edition. Revised and Enlarged. Medium 8vo. 21s.

Day.—ELECTRIC LIGHT ARITHMETIC. By R. E. DAY, M.A., Evening Lecturer in Experimental Physics at King's College, London. Pott 8vo. 2s.

Everett.—UNITS AND PHYSICAL CONSTANTS. By J. D. EVERETT, M.A., D.C.L., F.R.S., F.R.S.E., Professor of Natural Philosophy, Queen's College, Belfast. Second Edition. Extra fcap. 8vo. 5s.

Gray.—THE THEORY AND PRACTICE OF ABSOLUTE MEASUREMENTS IN ELECTRICITY AND MAGNETISM. By ANDREW GRAY, M.A., F.R.S.E., Professor of Physics in the University College of North Wales. Two Vols. Crown 8vo. Vol. I. 12s. 6d. [Vol. II. *In the press.*

Greaves.—STATICS FOR SCHOOLS. By JOHN GREAVES, M.A., Fellow and Mathematical Lecturer of Christ's College, Cambridge. [*In preparation.*

ELEMENTARY STATICS, A TREATISE ON. By the same. Second Edition, revised. Crown 8vo. 6s. 6d.

Grove.—A DICTIONARY OF MUSIC AND MUSICIANS. (A.D. 1450—1888). By Eminent Writers, English and Foreign. Edited by Sir GEORGE GROVE, D.C.L., Director of the Royal College of Music, &c. Demy 8vo.

Vols. I., II., and III. Price 21s. each.

Vol. I. A to IMPROMPTU. Vol. II. IMPROPERIA to PLAIN SONG. Vol. III. PLANCHÉ TO SUMER IS ICUMEN IN. Demy 8vo. cloth, with Illustrations in Music Type and Woodcut. Also published in Parts. Parts I. to XIV., Parts XIX—XXII., price 3s. 6d. each. Parts XV., XVI., price 7s. Parts XVII., XVIII., price 7s.

*** (Part XXII.) just published, completes the DICTIONARY OF MUSIC AND MUSICIANS as originally contemplated. But an Appendix and a full general Index are in the press.

"Dr. Grove's Dictionary will be a boon to every intelligent lover of music."—SATURDAY REVIEW.

Huxley.—INTRODUCTORY PRIMER OF SCIENCE. By T. H. HUXLEY, F.R.S., &c. 18mo. 1s.

Ibbetson.—THE MATHEMATICAL THEORY OF PERFECTLY ELASTIC SOLIDS, with a Short Account of Viscous Fluids. An Elementary Treatise. By WILLIAM JOHN IBBETSON, B.A., F.R.A.S., Senior Scholar of Clare College, Cambridge. 8vo. Price 21s.

Jones.—EXAMPLES IN PHYSICS. By D. E. JONES, B.Sc. Lecturer in Physics in University College of Wales, Aberystwyth. Fcap. 8vo. 3s. 6d.

Kempe.—HOW TO DRAW A STRAIGHT LINE; a Lecture on Linkages. By A. B. KEMPE. With Illustrations. Crown 8vo. 1s. 6d. (*Nature Series.*)

Kennedy.—THE MECHANICS OF MACHINERY. By A. B. W. KENNEDY, F.R.S., M.Inst.C.E., Professor of Engineering and Mechanical Technology in University College, London. With numerous Illustrations. Crown 8vo. 12s. 6d.

SCIENCE.

Lang.—EXPERIMENTAL PHYSICS. By P. R. SCOTT LANG, M.A., Professor of Mathematics in the University of St. Andrews. With Illustrations. Crown 8vo. [*In the press.*

Lock.—Works by Rev. J. B. LOCK, M.A., Senior Fellow, Assistant Tutor and Lecturer in Gonville and Caius College, Cambridge, formerly Master at Eton, &c.
DYNAMICS FOR BEGINNERS. Globe 8vo. 3s. 6d.
ELEMENTARY STATICS. Globe 8vo. 4s. 6d.

Lodge.—MODERN VIEWS OF ELECTRICITY. By OLIVER J. LODGE, F.R.S., Professor of Physics in University College, Liverpool. Illustrated. Crown 8vo. [*In preparation.*

Loewy.—QUESTIONS AND EXAMPLES ON EXPERIMENTAL PHYSICS: Sound, Light, Heat, Electricity, and Magnetism. By B. LOEWY, F.R.A.S., Science Master at the London International College, and Examiner in Experimental Physics to the College of Preceptors. Fcap. 8vo. 2s.

Lupton.—NUMERICAL TABLES AND CONSTANTS IN ELEMENTARY SCIENCE. By SYDNEY LUPTON, M.A,. F.C.S., F.I.C., Assistant Master at Harrow School. Extra fcap. 8vo. 2s. 6d.

Macfarlane.—PHYSICAL ARITHMETIC. By ALEXANDER MACFARLANE, D.Sc., Examiner in Mathematics in the University of Edinburgh. Crown 8vo. 7s. 6d.

Macgregor.—KINEMATICS AND DYNAMICS. An Elementary Treatise. By JAMES GORDON MACGREGOR, M.A., D. Sc., Fellow of the Royal Societies of Edinburgh and of Canada, Munro Professor of Physics in Dalhousie College, Halifax, Nova Scotia. With Illustrations. Crown 8vo. 10s. 6d.

Mayer.—SOUND: a Series of Simple, Entertaining, and Inexpensive Experiments in the Phenomena of Sound, for the Use of Students of every age. By A. M. MAYER, Professor of Physics in the Stevens Institute of Technology, &c. With numerous Illustrations. Crown 8vo. 2s. 6d. (*Nature Series.*)

Mayer and Barnard.—LIGHT: a Series of Simple, Entertaining, and Inexpensive Experiments in the Phenomena of Light, for the Use of Students of every age. By A. M. MAYER and C. BARNARD. With numerous Illustrations. Crown 8vo. 2s. 6d. (*Nature Series.*)

Newton.—PRINCIPIA. Edited by Professor Sir W. THOMSON and Professor BLACKBURNE. 4to, cloth. 31s. 6d.
THE FIRST THREE SECTIONS OF NEWTON'S PRINCIPIA. With Notes and Illustrations. Also a Collection of Problems, principally intended as Examples of Newton's Methods. By PERCIVAL FROST, M.A. Third Edition. 8vo. 12s.

Parkinson.—A TREATISE ON OPTICS. By S. PARKINSON, D.D., F.R.S., Tutor and Prælector of St. John's College, Cambridge. Fourth Edition, revised and enlarged. Crown 8vo. 10s. 6d.

Perry.—STEAM. AN ELEMENTARY TREATISE. By JOHN PERRY, C.E., Whitworth Scholar, Fellow of the Chemical Society, Professor of Mechanical Engineering and Applied Mechanics at the Technical College, Finsbury. With numerous Woodcuts and Numerical Examples and Exercises. 18mo. 4s. 6d.

Ramsay.—EXPERIMENTAL PROOFS OF CHEMICAL THEORY FOR BEGINNERS. By WILLIAM RAMSAY, Ph.D., Professor of Chemistry in University Coll., Bristol. Pott 8vo. 2s. 6d.

Rayleigh.—THE THEORY OF SOUND. By LORD RAYLEIGH, M.A., F.R.S., formerly Fellow of Trinity College, Cambridge, 8vo. Vol. I. 12s. 6d. Vol. II. 12s. 6d. [*Vol. III. in the press.*

Reuleaux.—THE KINEMATICS OF MACHINERY. Outlines of a Theory of Machines. By Professor F. REULEAUX. Translated and Edited by Professor A. B. W. KENNEDY, F.R.S., C.E. With 450 Illustrations. Medium 8vo. 21s.

Roscoe and Schuster.—SPECTRUM ANALYSIS. Lectures delivered in 1868 before the Society of Apothecaries of London. By Sir HENRY E. ROSCOE, LL.D., F.R.S., formerly Professor of Chemistry in the Owens College, Victoria University, Manchester. Fourth Edition, revised and considerably enlarged by the Author and by ARTHUR SCHUSTER, F.R.S., Ph.D., Professor of Applied Mathematics in the Owens College, Victoria University. With Appendices, numerous Illustrations, and Plates. Medium 8vo. 21s.

Shann.—AN ELEMENTARY TREATISE ON HEAT, IN RELATION TO STEAM AND THE STEAM-ENGINE. By G. SHANN, M.A. With Illustrations. Crown 8vo. 4s. 6d.

Spottiswoode.—POLARISATION OF LIGHT. By the late W. SPOTTISWOODE, F.R.S. With many Illustrations. New Edition. Crown 8vo. 3s. 6d. (*Nature Series.*)

Stewart (Balfour).—Works by BALFOUR STEWART, F.R.S., late Langworthy Professor of Physics in the Owens College, Victoria University, Manchester.
PRIMER OF PHYSICS. With numerous Illustrations. New Edition, with Questions. 18mo. 1s. (*Science Primers.*)
LESSONS IN ELEMENTARY PHYSICS. With numerous Illustrations and Chromolitho of the Spectra of the Sun, Stars, and Nebulæ. New and Enlarged Edition. Fcap. 8vo. 4s. 6d.
QUESTIONS ON BALFOUR STEWART'S ELEMENTARY LESSONS IN PHYSICS. By Prof. THOMAS H. CORE, Owens College, Manchester. Fcap. 8vo. 2s.

Stewart and Gee.—ELEMENTARY PRACTICAL PHYSICS, LESSONS IN. By BALFOUR STEWART, M.A., LL.D., F.R.S., and W. W. HALDANE GEE, B.Sc. Crown 8vo.
Vol. I.—GENERAL PHYSICAL PROCESSES. 6s.
Vol. II.—ELECTRICITY AND MAGNETISM. 7s. 6d.
Vol. III.—OPTICS, HEAT, AND SOUND. [*In the press.*
PRACTICAL PHYSICS FOR SCHOOLS AND THE JUNIOR STUDENTS OF COLLEGES. By the same Authors. Globe 8vo.
Vol. I.—ELECTRICITY AND MAGNETISM. 2s. 6d.

SCIENCE.

Stokes.—ON LIGHT. Being the Burnett Lectures, delivered in Aberdeen in 1883, 1884, 1885. By GEORGE GABRIEL STOKES, M.A., P.R.S., &c., Fellow of Pembroke College, and Lucasian Professor of Mathematics in the University of Cambridge. First Course: ON THE NATURE OF LIGHT.—Second Course: ON LIGHT AS A MEANS OF INVESTIGATION.—Third Course: ON THE BENEFICIAL EFFECTS OF LIGHT. Complete in one volume. Crown 8vo. 7s. 6d.

** The Second and Third Courses may be had separately. Crown 8vo. 2s. 6d. each.

Stone.—AN ELEMENTARY TREATISE ON SOUND. By W. H. STONE, M.D. With Illustrations. 18mo. 3s. 6d.

Tait.—HEAT. By P. G. TAIT, M.A., Sec. R.S.E., formerly Fellow of St. Peter's College, Cambridge, Professor of Natural Philosophy in the University of Edinburgh. Crown 8vo. 6s.

Thompson.—ELEMENTARY LESSONS IN ELECTRICITY AND MAGNETISM. By SILVANUS P. THOMPSON, Principal and Professor of Physics in the Technical College, Finsbury. With Illustrations. New Edition, Revised. Twenty-Eighth Thousand. Fcap. 8vo. 4s. 6d.

Thomson, Sir W.—ELECTROSTATICS AND MAGNETISM, REPRINTS OF PAPERS ON. By Sir WILLIAM THOMSON, D.C.L., LL.D, F.R.S., F.R.S.E., Fellow of St. Peter's College, Cambridge, and Professor of Natural Philosophy in the University of Glasgow. Second Edition. Medium 8vo. 18s.

Thomson, J. J.—Works by J. J. THOMSON, Fellow of Trinity College, Cavendish Professor of Experimental Physics in the University.

THE MOTION OF VORTEX RINGS, A TREATISE ON. An Essay to which the Adams Prize was adjudged in 1882 in the University of Cambridge. With Diagrams. 8vo. 6s.

APPLICATIONS OF DYNAMICS TO PHYSICS AND CHEMISTRY. By the same Author. Crown 8vo. 7s. 6d.

Todhunter.—NATURAL PHILOSOPHY FOR BEGINNERS. By I. TODHUNTER, M.A., F.R.S., D.Sc.
Part I. The Properties of Solid and Fluid Bodies. 18mo. 3s. 6d.
Part II. Sound, Light, and Heat. 18mo. 3s. 6d.

Turner.—HEAT AND ELECTRICITY, A COLLECTION OF EXAMPLES ON. By H. H. TURNER, B.A., Fellow of Trinity College, Cambridge. Crown 8vo. 2s. 6d.

Wright (Lewis).—LIGHT; A COURSE OF EXPERIMENTAL OPTICS, CHIEFLY WITH THE LANTERN. By LEWIS WRIGHT. With nearly 200 Engravings and Coloured Plates. Crown 8vo. 7s. 6d.

ASTRONOMY.

Airy.—POPULAR ASTRONOMY. With Illustrations by Sir G. B. AIRY, K.C.B., formerly Astronomer-Royal. New Edition. 18mo. 4s. 6d.

Forbes.—TRANSIT OF VENUS. By G. FORBES, M.A., Professor of Natural Philosophy in the Andersonian University, Glasgow. Illustrated. Crown 8vo. 3s. 6d. (*Nature Series.*)

Godfray.—Works by HUGH GODFRAY, M.A., Mathematical Lecturer at Pembroke College, Cambridge.

A TREATISE ON ASTRONOMY, for the Use of Colleges and Schools. Fourth Edition. 8vo. 12s. 6d.

AN ELEMENTARY TREATISE ON THE LUNAR THEORY, with a Brief Sketch of the Problem up to the time of Newton. Second Edition, revised. Crown 8vo. 5s. 6d.

Lockyer.—Works by J. NORMAN LOCKYER, F.R.S.

PRIMER OF ASTRONOMY. With numerous Illustrations. New Edition. 18mo. 1s. (*Science Primers.*)

ELEMENTARY LESSONS IN ASTRONOMY. With Coloured Diagram of the Spectra of the Sun, Stars, and Nebulæ, and numerous Illustrations. New Edition, revised. Fcap. 8vo. 5s. 6d.

QUESTIONS ON LOCKYER'S ELEMENTARY LESSONS IN ASTRONOMY. For the Use of Schools. By JOHN FORBES-ROBERTSON. 18mo, cloth limp. 1s. 6d.

THE CHEMISTRY OF THE SUN. With Illustrations. 8vo. 14s.

Newcomb.—POPULAR ASTRONOMY. By S. NEWCOMB, LL.D., Professor U.S. Naval Observatory. With 112 Illustrations and 5 Maps of the Stars. Second Edition, revised. 8vo. 18s.

"It is unlike anything else of its kind, and will be of more use in circulating a knowledge of Astronomy than nine-tenths of the books which have appeared on the subject of late years."—SATURDAY REVIEW.

CHEMISTRY.

Armstrong.—A MANUAL OF INORGANIC CHEMISTRY. By HENRY ARMSTRONG, Ph.D., F.R.S., Professor of Chemistry in the City and Guilds of London Technical Institute. Crown 8vo. [*In preparation.*

Cohen.—THE OWENS COLLEGE COURSE OF PRACTICAL ORGANIC CHEMISTRY. By JULIUS B. COHEN, Ph.D., F.C.S., Assistant Lecturer on Chemistry in the Owens College, Manchester. With a Preface by SIR HENRY ROSCOE, F.R.S., and C. SCHORLEMMER, F.R.S. Fcap. 8vo. 2s. 6d.

Cooke.—ELEMENTS OF CHEMICAL PHYSICS. By JOSIAH P. COOKE, Junr., Erving Professor of Chemistry and Mineralogy in Harvard University. Fourth Edition. Royal 8vo. 21s.

SCIENCE.

Fleischer.—A SYSTEM OF VOLUMETRIC ANALYSIS. By EMIL FLEISCHER. Translated, with Notes and Additions, from the Second German Edition by M. M. PATTISON MUIR, F.R.S.E. With Illustrations. Crown 8vo. 7s. 6d.

Frankland.—AGRICULTURAL CHEMICAL ANALYSIS A Handbook of. By PERCY FARADAY FRANKLAND, Ph.D., B.Sc., F.C.S. Associate of the Royal School of Mines, and Demonstrator of Practical and Agricultural Chemistry in the Normal School of Science and Royal School of Mines, South Kensington Museum. Founded upon *Leitfaden für die Agriculture Chemiche Analyse*, von Dr. F. KROCKER. Crown 8vo. 7s. 6d.

Hartley.—A COURSE OF QUANTITATIVE ANALYSIS FOR STUDENTS. By W. NOEL HARTLEY, F.R.S., Professor of Chemistry, and of Applied Chemistry, Science and Art Department, Royal College of Science, Dublin. Globe 8vo. 5s.

Hiorns.—Works by ARTHUR H. HIORNS, Principal of the School of Metallurgy, Birmingham and Midland Institute.
PRACTICAL METALLURGY AND ASSAYING. A Text-Book for the use of Teachers, Students, and Assayers. With Illustrations. Globe 8vo. 6s.
A TEXT-BOOK ON ELEMENTARY THEORETICAL METALLURGY. Globe 8vo. [*In the press.*

Jones.—Works by FRANCIS JONES, F.R.S.E., F.C.S., Chemical Master in the Grammar School, Manchester.
THE OWENS COLLEGE JUNIOR COURSE OF PRACTICAL CHEMISTRY. With Preface by Sir HENRY ROSCOE, F.R.S., and Illustrations. New Edition. 18mo. 2s. 6d.
QUESTIONS ON CHEMISTRY. A Series of Problems and Exercises in Inorganic and Organic Chemistry. Fcap. 8vo. 3s.

Landauer.—BLOWPIPE ANALYSIS. By J. LANDAUER. Authorised English Edition by J. TAYLOR and W. E. KAY, of Owens College, Manchester. Extra fcap. 8vo. 4s. 6d.

Lupton.—CHEMICAL ARITHMETIC. With 1,200 Problems. By SYDNEY LUPTON, M.A., F.C.S., F.I.C., formerly Assistant-Master at Harrow. Second Edition, Revised and Abridged. Fcap. 8vo. 4s. 6d.

Meldola.—PHOTOGRAPHIC CHEMISTRY. By RAPHAEL MELDOLA, F.R.S., Professor of Chemistry in the Technical College, Finsbury. Crown 8vo. (*Nature Series.*) [*In the press.*

Muir.—PRACTICAL CHEMISTRY FOR MEDICAL STUDENTS. Specially arranged for the first M.B. Course. By M. M. PATTISON MUIR, F.R.S.E. Fcap. 8vo. 1s. 6d.

Muir and Wilson.—THE ELEMENTS OF THERMAL CHEMISTRY. By M. M. PATTISON MUIR, M.A., F.R.S.E., Fellow and Prælector of Chemistry in Gonville and Caius Colleges, Cambridge; Assisted by DAVID MUIR WILSON. 8vo. 12s. 6d.

Remsen.—Works by IRA REMSEN, Professor of Chemistry in the Johns Hopkins University.

    COMPOUNDS OF CARBON; or, Organic Chemistry, an Introduction to the Study of. Crown 8vo. 6s. 6d.

    AN INTRODUCTION TO THE STUDY OF CHEMISTRY (INORGANIC CHEMISTRY). Crown 8vo. 6s. 6d.

    THE ELEMENTS OF CHEMISTRY. A Text Book for Beginners. Fcap. 8vo. 2s. 6d.

Roscoe.—Works by Sir HENRY E. ROSCOE, F.R.S., formerly Professor of Chemistry in the Victoria University the Owens College, Manchester.

    PRIMER OF CHEMISTRY. With numerous Illustrations. New Edition. With Questions. 18mo. 1s. (*Science Primers.*)

    LESSONS IN ELEMENTARY CHEMISTRY, INORGANIC AND ORGANIC. With numerous Illustrations and Chromolitho of the Solar Spectrum, and of the Alkalies and Alkaline Earths. New Edition. Fcap. 8vo. 4s. 6d. (*See under* THORPE.)

Roscoe and Schorlemmer.—INORGANIC AND ORGANIC CHEMISTRY. A Complete Treatise on Inorganic and Organic Chemistry. By Sir HENRY E. ROSCOE, F.R.S., and Prof. C. SCHORLEMMER, F.R.S. With Illustrations. Medium 8vo.

Vols. I. and II.—INORGANIC CHEMISTRY.

    Vol. I.—The Non-Metallic Elements. Second Edition, revised. 21s. Vol. II. Part I.—Metals. 18s. Vol. II. Part II.—Metals. 18s.

Vol. III.—ORGANIC CHEMISTRY.

    THE CHEMISTRY OF THE HYDROCARBONS and their Derivatives, or ORGANIC CHEMISTRY. With numerous Illustrations. Five Parts. Parts I., II., and IV. 21s. each. Part III. 18s. [Part V. *Immediately.*

Thorpe.—A SERIES OF CHEMICAL PROBLEMS, prepared with Special Reference to Sir H. E. Roscoe's Lessons in Elementary Chemistry, by T. E. THORPE, Ph.D., F.R.S., Professor of Chemistry in the Normal School of Science, South Kensington, adapted for the Preparation of Students for the Government, Science, and Society of Arts Examinations. With a Preface by Sir HENRY E. ROSCOE, F.R.S. New Edition, with Key. 18mo. 2s.

Thorpe and Rücker.—A TREATISE ON CHEMICAL PHYSICS. By T. E. THORPE, Ph.D., F.R.S. Professor of Chemistry in the Normal School of Science, and Professor A. W. RÜCKER. Illustrated. 8vo. [*In preparation.*

Wright.—METALS AND THEIR CHIEF INDUSTRIAL APPLICATIONS. By C. ALDER WRIGHT, D.Sc., &c., Lecturer on Chemistry in St. Mary's Hospital Medical School. Extra fcap. 8vo. 3s. 6d.

BIOLOGY.

Allen.—ON THE COLOUR OF FLOWERS, as Illustrated in the British Flora. By GRANT ALLEN. With Illustrations. Crown 8vo. 3s. 6d. (*Nature Series.*)

Balfour.— A TREATISE ON COMPARATIVE EMBRYOLOGY. By F. M. BALFOUR, M.A., F.R.S., Fellow and Lecturer of Trinity College, Cambridge. With Illustrations. Second Edition, reprinted without alteration from the First Edition. In 2 vols. 8vo. Vol. I. 18s. Vol. II. 21s.

Balfour and Ward.—A GENERAL TEXT BOOK OF BOTANY. By ISAAC BAYLEY BALFOUR, F.R.S., Professor of Botany in the University of Edinburgh, and H. MARSHALL WARD, F.R.S., Fellow of Christ's College, Cambridge, and Professor of Botany in the Royal Indian Engineering College, Cooper's Hill. 8vo. [*In preparation.*

Bettany.—FIRST LESSONS IN PRACTICAL BOTANY. By G. T. BETTANY, M.A., F.L.S., formerly Lecturer in Botany at Guy's Hospital Medical School. 18mo. 1s.

Bower.—A COURSE OF PRACTICAL INSTRUCTION IN BOTANY. By F. O. BOWER, D.Sc., F.L.S., Regius Professor of Botany in the University of Glasgow. Crown 8vo. Part I. Second Edition, revised and enlarged. Phanerogamæ—Pteridophyta. 6s. 6d. Part II. Bryophyta—Thallophyta. 4s. 6d. Or both Parts in one volume, 10s. 6d.

Darwin (Charles).—MEMORIAL NOTICES OF CHARLES DARWIN, F.R.S., &c. By THOMAS HENRY HUXLEY, F.R.S., G. J. ROMANES, F.R.S., ARCHIBALD GEIKIE, F.R.S., and W. T. THISELTON DYER, F.R.S. Reprinted from *Nature*. With a Portrait, engraved by C. H. JEENS. Crown 8vo. 2s. 6d. (*Nature Series.*)

Fearnley.—A MANUAL OF ELEMENTARY PRACTICAL HISTOLOGY. By WILLIAM FEARNLEY. With Illustrations. Crown 8vo. 7s. 6d.

Flower and Gadow.—AN INTRODUCTION TO THE OSTEOLOGY OF THE MAMMALIA. By WILLIAM HENRY FLOWER, LL.D., F.R.S., Director of the Natural History Departments of the British Museum, late Hunterian Professor of Comparative Anatomy and Physiology in the Royal College of Surgeons of England. With numerous Illustrations. Third Edition. Revised with the assistance of HANS GADOW, Ph.D., M.A., Lecturer on the Advanced Morphology of Vertebrates and Strickland Curator in the University of Cambridge. Crown 8vo. 10s. 6d.

Foster.—Works by MICHAEL FOSTER, M.D., Sec. R.S., Professor of Physiology in the University of Cambridge.
PRIMER OF PHYSIOLOGY. With numerous Illustrations. New Edition. 18mo. 1s.

Foster—*continued.*
A TEXT-BOOK OF PHYSIOLOGY. With Illustrations. Fifth Edition, largely revised. In Three Parts. Part I., comprising Book I. Blood—The Tissues of Movement, The Vascular Mechanism. 10s. 6d.
Parts II. and III. are in the press preparing for early publication.

Foster and Balfour.—THE ELEMENTS OF EMBRYOLOGY. By MICHAEL FOSTER, M.A., M.D., LL.D., Sec. R.S., Professor of Physiology in the University of Cambridge, Fellow of Trinity College, Cambridge, and the late FRANCIS M. BALFOUR, M.A., LL.D., F.R.S., Fellow of Trinity College, Cambridge, and Professor of Animal Morphology in the University. Second Edition, revised. Edited by ADAM SEDGWICK, M.A., Fellow and Assistant Lecturer of Trinity College, Cambridge, and WALTER HEAPE, Demonstrator in the Morphological Laboratory of the University of Cambridge. With Illustrations. Crown 8vo. 10s. 6d.

Foster and Langley.—A COURSE OF ELEMENTARY PRACTICAL PHYSIOLOGY. By Prof. MICHAEL FOSTER, M.D., Sec. R.S., &c., and J. N. LANGLEY, M.A., F.R.S., Fellow of Trinity College, Cambridge. Fifth Edition. Crown 8vo. 7s. 6d.

Gamgee.—A TEXT-BOOK OF THE PHYSIOLOGICAL CHEMISTRY OF THE ANIMAL BODY. Including an Account of the Chemical Changes occurring in Disease. By A. GAMGEE, M.D., F.R.S., formerly Professor of Physiology in the Victoria University the Owens College, Manchester. 2 Vols. 8vo. With Illustrations. Vol. I. 18s. [*Vol. II. in the press.*

Gray.—STRUCTURAL BOTANY, OR ORGANOGRAPHY ON THE BASIS OF MORPHOLOGY. To which are added the principles of Taxonomy and Phytography, and a Glossary of Botanical Terms. By Professor ASA GRAY, LL.D. 8vo. 10s. 6d.

Hamilton.—A PRACTICAL TEXT-BOOK OF PATHOLOGY. By D. J. HAMILTON, Professor of Pathological Anatomy (Sir Erasmus Wilson Chair), University of Aberdeen. Illustrated. 8vo. [*In the press.*

Hooker.—Works by Sir J. D. HOOKER, K.C.S.I., C.B., M.D., F.R.S., D.C.L.
PRIMER OF BOTANY. With numerous Illustrations. New Edition. 18mo. 1s. (*Science Primers.*)
THE STUDENT'S FLORA OF THE BRITISH ISLANDS. Third Edition, revised. Globe 8vo. 10s. 6d.

Howes.—AN ATLAS OF PRACTICAL ELEMENTARY BIOLOGY. By G. B. HOWES, Assistant Professor of Zoology, Normal School of Science and Royal School of Mines. With a Preface by THOMAS HENRY HUXLEY, F.R.S. Royal 4to. 14s.

SCIENCE.

Huxley.—Works by THOMAS HENRY HUXLEY, F.R.S.
INTRODUCTORY PRIMER OF SCIENCE. 18mo. 1s.
(*Science Primers.*)
LESSONS IN ELEMENTARY PHYSIOLOGY. With numerous Illustrations. New Edition Revised. Fcap. 8vo. 4s. 6d.
QUESTIONS ON HUXLEY'S PHYSIOLOGY FOR SCHOOLS. By T. ALCOCK, M.D. New Edition. 18mo. 1s. 6d.

Huxley and Martin.—A COURSE OF PRACTICAL INSTRUCTION IN ELEMENTARY BIOLOGY. By T. H. HUXLEY, F.R.S., LL.D., assisted by H. N. MARTIN, M.A., M.B., D.Sc., F.R.S., Fellow of Christ's College, Cambridge. New Edition, revised and extended by G. B. HOWES, Assistant Professor of Zoology, Normal School of Science, and Royal School of Mines, and D. H. SCOTT, M.A., PH.D., Assistant Professor of Botany, Normal School of Science, and Royal School of Mines. New Edition, thoroughly revised. With a Preface by T. H. HUXLEY, F.R.S. Crown 8vo. 10s. 6d.

Kane.—EUROPEAN BUTTERFLIES, A HANDBOOK OF. By W. F. DE VISMES KANE, M.A., M.R.I.A., Member of the Entomological Society of London, &c. With Copper Plate Illustrations. Crown 8vo. 10s. 6d.
A LIST OF EUROPEAN RHOPALOCERA WITH THEIR VARIETIES AND PRINCIPAL SYNONYMS. Reprinted from the *Handbook of European Butterflies*. Crown 8vo. 1s.

Klein.—MICRO-ORGANISMS AND DISEASE. An Introduction into the Study of Specific Micro-Organisms. By E. KLEIN, M.D., F.R.S,, Lecturer on General Anatomy and Physiology in the Medical School of St. Bartholomew's Hospital, London. With 121 Illustrations. Third Edition, Revised. Crown 8vo. 6s.
THE BACTERIA IN ASIATIC CHOLERA. By the Same. Crown 8vo. [*In preparation.*

Lankester.—Works by Professor E. RAY LANKESTER, F.R.S.
A TEXT BOOK OF ZOOLOGY. 8vo. [*In preparation.*
DEGENERATION: A CHAPTER IN DARWINISM. Illustrated. Crown 8vo. 2s. 6d. (*Nature Series.*)

Lubbock.—Works by SIR JOHN LUBBOCK, M.P., F.R.S., D.C.L.
THE ORIGIN AND METAMORPHOSES OF INSECTS. With numerous Illustrations. New Edition. Crown 8vo. 3s. 6d. (*Nature Series.*)
ON BRITISH WILD FLOWERS CONSIDERED IN RELATION TO INSECTS. With numerous Illustrations. New Edition. Crown 8vo. 4s. 6d. (*Nature Series.*)
FLOWERS, FRUITS, AND LEAVES. With Illustrations. Second Edition. Crown 8vo. 4s. 6d. (*Nature Series.*)

Martin and Moale.—ON THE DISSECTION OF VERTEBRATE ANIMALS. By Professor H. N. MARTIN and W. A. MOALE. Crown 8vo. [*In preparation.*

Mivart.—Works by ST. GEORGE MIVART, F.R.S., Lecturer on Comparative Anatomy at St. Mary's Hospital.
   LESSONS IN ELEMENTARY ANATOMY. With upwards of 400 Illustrations. Fcap. 8vo. 6s. 6d.
   THE COMMON FROG. Illustrated. Cr. 8vo. 3s. 6d. (*Nature Series.*)

Müller.—THE FERTILISATION OF FLOWERS. By Professor HERMANN MÜLLER. Translated and Edited by D'ARCY W. THOMPSON, B.A., Professor of Biology in University College, Dundee. With a Preface by CHARLES DARWIN, F.R.S. With numerous Illustrations. Medium 8vo. 21s.

Oliver.—Works by DANIEL OLIVER, F.R.S., &c., Professor of Botany in University College, London, &c.
   FIRST BOOK OF INDIAN BOTANY. With numerous Illustrations. Extra fcap. 8vo. 6s. 6d.
   LESSONS IN ELEMENTARY BOTANY. With nearly 200 Illustrations. New Edition. Fcap. 8vo. 4s. 6d.

Parker.—A COURSE OF INSTRUCTION IN ZOOTOMY (VERTEBRATA). By T. JEFFREY PARKER, B.Sc. London, Professor of Biology in the University of Otago, New Zealand. With Illustrations. Crown 8vo. 8s. 6d.
   LESSONS IN ELEMENTARY BIOLOGY. By the same Author. With Illustrations. 8vo. [*In the press.*

Parker and Bettany.—THE MORPHOLOGY OF THE SKULL. By Professor W. K. PARKER, F.R.S., and G. T. BETTANY. Illustrated. Crown 8vo. 10s. 6d.

Romanes.—THE SCIENTIFIC EVIDENCES OF ORGANIC EVOLUTION. By GEORGE J. ROMANES, M.A., LL D., F.R.S., Zoological Secretary of the Linnean Society. Crown 8vo. 2s. 6d. (*Nature Series.*)

Sedgwick.—A SUPPLEMENT TO F. M. BALFOUR'S TREATISE ON EMBRYOLOGY. By ADAM SEDGWICK, M.A., F.R.S., Fellow and Lecturer of Trinity College, Cambridge. 8vo. Illustrated. [*In preparation.*

Smith (W. G.).—DISEASES OF FIELD AND GARDEN CROPS, CHIEFLY SUCH AS ARE CAUSED BY FUNGI. By WORTHINGTON G. SMITH, F.L.S., M.A.I., Member of the Scientific Committee R.H.S. With 143 New Illustrations drawn and engraved from Nature by the Author. Fcap. 8vo. 4s. 6d.

Stewart—Corry.—A FLORA OF THE NORTH-EAST OF IRELAND. Including the Phanerogamia, the Cryptogamia Vascularia, and the Muscineæ. By SAMUEL ALEXANDER STEWART, Fellow of the Botanical Society of Edinburgh, Curator of the Collections in the Belfast Museum, and Honorary Associate of the Belfast Natural History and Philosophical Society; and the late THOMAS HUGHES CORRY, M.A., F.L.S., F.Z.S., M.R.I.A., F.B.S. Edin., Lecturer on Botany in the University Medical and Science Schools, Cambridge, Assistant Curator of the University Herbarium, &c., &c. Crown 8vo. 5s. 6d.

SCIENCE.

Ward.—TIMBER AND ITS DISEASES. By H. MARSHALL WARD, F.R.S., Professor of Botany in the Royal Indian Engineering College, Cooper's Hill. Illustrated. Crown 8vo.
[In preparation.

Wiedersheim (Prof.).—ELEMENTS OF THE COMPARATIVE ANATOMY OF VERTEBRATES. Adapted from the German of ROBERT WIEDERSHEIM, Professor of Anatomy, and Director of the Institute of Human and Comparative Anatomy in the University of Freiburg-in-Baden, by W. NEWTON PARKER, Professor of Biology in the University College of South Wales and Monmouthshire. With Additions by the Author and Translator. With Two Hundred and Seventy Woodcuts. Medium 8vo. 12s. 6d.

MEDICINE.

Brunton.—Works by T. LAUDER BRUNTON, M.D., D.Sc., F.R.C.P., F.R.S., Assistant Physician and Lecturer on Materia Medica at St. Bartholomew's Hospital; Examiner in Materia Medica in the University of London, in the Victoria University, and in the Royal College of Physicians, London; late Examiner in the University of Edinburgh.

A TEXT-BOOK OF PHARMACOLOGY, THERAPEUTICS, AND MATERIA MEDICA. Adapted to the United States Pharmacopœia, by FRANCIS H. WILLIAMS, M.D., Boston, Mass. Third Edition. Adapted to the New British Pharmacopœia, 1885. Medium 8vo. 21s.

TABLES OF MATERIA MEDICA: A Companion to the Materia Medica Museum. With Illustrations. New Edition Enlarged. 8vo. 10s. 6d.

Griffiths.—LESSONS ON PRESCRIPTIONS AND THE ART OF PRESCRIBING. By W. HANDSEL GRIFFITHS, PH.D., L.R.C.P.E. New Edition. Adapted to the Pharmacopœia, 1885. 18mo. 3s. 6d.

Hamilton.—A TEXT-BOOK OF PATHOLOGY. By D. J. HAMILTON, Professor of Pathological Anatomy (Sir Erasmus Wilson Chair) University of Aberdeen. With Illustrations. 8vo.
[In the press.

Klein.—MICRO-ORGANISMS AND DISEASE. An Introduction into the Study of Specific Micro-Organisms. By E. KLEIN, M.D., F.R.S., Lecturer on General Anatomy and Physiology in the Medical School of St. Bartholomew's Hospital, London. With 121 Illustrations. Third Edition, Revised. Crown 8vo. 6s.

THE BACTERIA IN ASIATIC CHOLERA. By the Same Author. Crown 8vo. *[In preparation.*

Ziegler-Macalister.—TEXT-BOOK OF PATHOLOGICAL ANATOMY AND PATHOGENESIS. By Professor ERNST ZIEGLER of Tübingen. Translated and Edited for English Students by DONALD MACALISTER, M.A., M.D., B.Sc., F.R.C.P., Fellow and Medical Lecturer of St. John's College, Cambridge, Physician to Addenbrooke's Hospital, and Teacher of Medicine in the University. With numerous Illustrations. Medium 8vo.
  Part I.—GENERAL PATHOLOGICAL ANATOMY. Second Edition. 12s. 6d.
  Part II.—SPECIAL PATHOLOGICAL ANATOMY. Sections I.—VIII. Second Edition. 12s. 6d. Sections IX.—XII. 12s. 6d.

ANTHROPOLOGY.

Flower.—FASHION IN DEFORMITY, as Illustrated in the Customs of Barbarous and Civilised Races. By Professor FLOWER, F.R.S., F.R.C.S. With Illustrations. Crown 8vo. 2s. 6d. (*Nature Series.*)

Tylor.—ANTHROPOLOGY. An Introduction to the Study of Man and Civilisation. By E. B. TYLOR, D.C.L., F.R.S. With numerous Illustrations. Crown 8vo. 7s. 6d.

PHYSICAL GEOGRAPHY & GEOLOGY.

Blanford.—THE RUDIMENTS OF PHYSICAL GEOGRAPHY FOR THE USE OF INDIAN SCHOOLS; with a Glossary of Technical Terms employed. By H. F. BLANFORD, F.R.S. New Edition, with Illustrations. Globe 8vo. 2s. 6d.

Geikie.—Works by ARCHIBALD GEIKIE, LL.D., F.R.S., Director General of the Geological Survey of Great Britain and Ireland, and Director of the Museum of Practical Geology, London, formerly Murchison Professor of Geology and Mineralogy in the University of Edinburgh, &c.
  PRIMER OF PHYSICAL GEOGRAPHY. With numerous Illustrations. New Edition. With Questions. 18mo. 1s. (*Science Primers.*)
  ELEMENTARY LESSONS IN PHYSICAL GEOGRAPHY. With numerous Illustrations. New Edition. Fcap. 8vo. 4s. 6d.
  QUESTIONS ON THE SAME. 1s. 6d.
  PRIMER OF GEOLOGY. With numerous Illustrations. New Edition. 18mo. 1s. (*Science Primers.*)
  CLASS BOOK OF GEOLOGY. With upwards of 200 New Illustrations. Crown 8vo. 10s. 6d.
  TEXT-BOOK OF GEOLOGY. With numerous Illustrations. Second Edition, Sixth Thousand, Revised and Enlarged. 8vo. 28s.

Geikie—*continued.*
> OUTLINES OF FIELD GEOLOGY. With Illustrations. New Edition. Extra fcap. 8vo. 3s. 6d.
>
> THE SCENERY AND GEOLOGY OF SCOTLAND, VIEWED IN CONNEXION WITH ITS PHYSICAL GEOLOGY. With numerous Illustrations. Crown 8vo. 12s. 6d. (See also under *Geography*.)

Huxley.—PHYSIOGRAPHY. An Introduction to the Study of Nature. By THOMAS HENRY HUXLEY, F.R.S. With numerous Illustrations, and Coloured Plates. New and Cheaper Edition. Crown 8vo. 6s.

Lockyer.—OUTLINES OF PHYSIOGRAPHY—THE MOVEMENTS OF THE EARTH. By J. NORMAN LOCKYER, F.R.S., Correspondent of the Institute of France, Foreign Member of the Academy of the Lyncei of Rome, &c., &c.; Professor of Astronomical Physics in the Normal School of Science, and Examiner in Physiography for the Science and Art Department. With Illustrations. Crown 8vo. Sewed, 1s. 6d.

Phillips.—A TREATISE ON ORE DEPOSITS. By J. ARTHUR PHILLIPS, F.R.S., V.P.G.S., F.C.S., M.Inst.C.E., Ancien Élève de l'École des Mines, Paris; Author of "A Manual of Metallurgy," "The Mining and Metallurgy of Gold and Silver," &c. With numerous Illustrations. 8vo. 25s.

AGRICULTURE.

Frankland.—AGRICULTURAL CHEMICAL ANALYSIS, A Handbook of. By PERCY FARADAY FRANKLAND, Ph.D., B.Sc., F.C.S., Associate of the Royal School of Mines, and Demonstrator of Practical and Agricultural Chemistry in the Normal School of Science and Royal School of Mines, South Kensington Museum. Founded upon *Leitfaden für die Agriculture Chemiche Analyse*, von Dr. F. KROCKER. Crown 8vo. 7s. 6d.

Smith (Worthington G.).—DISEASES OF FIELD AND GARDEN CROPS, CHIEFLY SUCH AS ARE CAUSED BY FUNGI. By WORTHINGTON G. SMITH, F.L.S., M.A.I., Member of the Scientific Committee of the R.H.S. With 143 Illustrations, drawn and engraved from Nature by the Author. Fcap. 8vo. 4s. 6d.

Tanner.—Works by HENRY TANNER, F.C.S., M.R.A.C., Examiner in the Principles of Agriculture under the Government Department of Science; Director of Education in the Institute of Agriculture, South Kensington, London; sometime Professor of Agricultural Science, University College, Aberystwith.
> ELEMENTARY LESSONS IN THE SCIENCE OF AGRICULTURAL PRACTICE. Fcap. 8vo. 3s. 6d.

Tanner—*continued.*
FIRST PRINCIPLES OF AGRICULTURE. 18mo. 1*s.*
THE PRINCIPLES OF AGRICULTURE. A Series of Reading Books for use in Elementary Schools. Prepared by HENRY TANNER, F.C.S., M.R.A.C. Extra fcap. 8vo.
   I. The Alphabet of the Principles of Agriculture. 6*d.*
   II. Further Steps in the Principles of Agriculture. 1*s.*
   III. Elementary School Readings on the Principles of Agriculture for the third stage. 1*s.*

POLITICAL ECONOMY.

Cairnes.—THE CHARACTER AND LOGICAL METHOD OF POLITICAL ECONOMY. By J. E. CAIRNES, LL.D, Emeritus Professor of Political Economy in University College, London. New Edition. Crown 8vo. 6*s.*

Cossa.—GUIDE TO THE STUDY OF POLITICAL ECONOMY. By Dr. LUIGI COSSA, Professor in the University of Pavia. Translated from the Second Italian Edition. With a Preface by W. STANLEY JEVONS, F.R.S. Crown 8vo. 4*s.* 6*d.*

Fawcett (Mrs.).—Works by MILLICENT GARRETT FAWCETT:—
POLITICAL ECONOMY FOR BEGINNERS, WITH QUESTIONS. Fourth Edition. 18mo. 2*s.* 6*d.*
TALES IN POLITICAL ECONOMY. Crown 8vo. 3*s.*

Fawcett.—A MANUAL OF POLITICAL ECONOMY. By Right Hon. HENRY FAWCETT, F.R.S. Sixth Edition, revised, with a chapter on "State Socialism and the Nationalisation of the Land," and an Index. Crown 8vo. 12*s.*
AN EXPLANATORY DIGEST of the above. By CYRIL A. WATERS, B.A. Crown 8vo. 2*s.* 6*d.*

Gunton.—WEALTH AND PROGRESS: A CRITICAL EXAMINATION OF THE WAGES QUESTION AND ITS ECONOMIC RELATION TO SOCIAL REFORM. By GEORGE GUNTON. Crown 8vo. 6*s.*

Jevons.—Works by W. STANLEY JEVONS, LL.D. (Edinb.), M.A. (Lond.), F.R.S., late Professor of Political Economy in University College, London, Examiner in Mental and Moral Science in the University of London.
PRIMER OF POLITICAL ECONOMY. New Edition. 18mo. 1*s.* (*Science Primers.*)
THE THEORY OF POLITICAL ECONOMY. Third Edition. Revised. Demy 8vo. 10*s.* 6*d.*

Marshall.—THE ECONOMICS OF INDUSTRY. By A. MARSHALL, M.A., Professor of Political Economy in the University of Cambridge, and MARY P. MARSHALL, late Lecturer at Newnham Hall, Cambridge. Extra fcap. 8vo. 2*s.* 6*d.*

Marshall.—ECONOMICS. By ALFRED MARSHALL, M.A., Professor of Political Economy in the University of Cambridge. 2 vols. 8vo.   [*In the press.*

SCIENCE.

Sidgwick.—THE PRINCIPLES OF POLITICAL ECONOMY. By Professor HENRY SIDGWICK, M.A., LL.D., Knightbridge Professor of Moral Philosophy in the University of Cambridge, &c., Author of "The Methods of Ethics." Second Edition, revised. 8vo. 16s.

Walker.—Works by FRANCIS A. WALKER, M.A., Ph.D., Author of "Money," "Money in its Relation to Trade," &c.
POLITICAL ECONOMY. Second Edition, revised and enlarged. 8vo. 12s. 6d.
A BRIEF TEXT-BOOK OF POLITICAL ECONOMY. Crown 8vo. 6s. 6d.
THE WAGES QUESTION. 8vo. 14s.

Wicksteed.—ALPHABET OF ECONOMIC SCIENCE. By PHILIP H. WICKSTEED, M.A. Globe 8vo. [*Just ready.*

MENTAL & MORAL PHILOSOPHY.

Boole.—THE MATHEMATICAL ANALYSIS OF LOGIC. Being an Essay towards a Calculus of Deductive Reasoning. By GEORGE BOOLE. 8vo. Sewed. 5s.

Calderwood.—HANDBOOK OF MORAL PHILOSOPHY. By the Rev. HENRY CALDERWOOD, LL.D., Professor of Moral Philosophy, University of Edinburgh. Fourteenth Edition, largely rewritten. Crown 8vo. 6s.

Clifford.—SEEING AND THINKING. By the late Professor W. K. CLIFFORD, F.R.S. With Diagrams. Crown 8vo. 3s. 6d. (*Nature Series.*)

Jardine.—THE ELEMENTS OF THE PSYCHOLOGY OF COGNITION. By the Rev. ROBERT JARDINE, B.D., D.Sc. (Edin.), Ex-Principal of the General Assembly's College, Calcutta. Third Edition, revised and improved. Crown 8vo. 6s. 6d.

Jevons.—Works by the late W. STANLEY JEVONS, LL.D., M.A., F.R.S.
PRIMER OF LOGIC. New Edition. 18mo. 1s. (*Science Primers.*)
ELEMENTARY LESSONS IN LOGIC; Deductive and Inductive, with copious Questions and Examples, and a Vocabulary of Logical Terms. New Edition. Fcap. 8vo. 3s. 6d.
THE PRINCIPLES OF SCIENCE. A Treatise on Logic and Scientific Method. New and Revised Edition. Crown 8vo. 12s. 6d.
STUDIES IN DEDUCTIVE LOGIC. Second Edition. Cr. 8vo. 6s.

Keynes.—FORMAL LOGIC, Studies and Exercises in. Including a Generalisation of Logical Processes in their application to Complex Inferences. By JOHN NEVILLE KEYNES, M.A., late Fellow of Pembroke College, Cambridge. Second Edition, Revised and Enlarged. Crown 8vo. 10s. 6d.

Kant—Max Müller.—CRITIQUE OF PURE REASON. By IMMANUEL KANT. In commemoration of the Centenary of its first Publication. Translated into English by F. MAX MÜLLER. With an Historical Introduction by LUDWIG NOIRÉ. 2 vols 8vo. 16s. each.
  Volume I. HISTORICAL INTRODUCTION, by LUDWIG NOIRÉ; &c., &c.
  Volume II. CRITIQUE OF PURE REASON, translated by F. MAX MÜLLER.
For the convenience of students these volumes are now sold separately.

Kant—Mahaffy and Bernard.—COMMENTARY ON KANT'S CRITIQUE. By J. P. MAHAFFY, M.A., Professor of Ancient History in the University of Dublin, and J. H. BERNARD, M.A. New and completed Edition. Crown 8vo. [*In preparation*.

McCosh.—PSYCHOLOGY. By JAMES McCOSH, D.D., LL.D., Litt.D. President of Princeton College, Author of "Intuitions of the Mind," "Laws of Discursive Thought," &c. Crown 8vo.
  I. THE COGNITIVE POWERS. 6s. 6d.
  II. THE MOTIVE POWERS. Crown 8vo. 6s. 6d.

Ray.—A TEXT-BOOK OF DEDUCTIVE LOGIC FOR THE USE OF STUDENTS. By P. K. RAY, D.Sc. (Lon. and Edin.), Professor of Logic and Philosophy, Presidency College Calcutta. Fourth Edition. Globe 8vo. 4s. 6d.
  The SCHOOLMASTER says:—"This work . . . is deservedly taking a place among the recognised text-books on Logic."

Sidgwick.—Works by HENRY SIDGWICK, M.A., LL.D., Knightbridge Professor of Moral Philosophy in the University of Cambridge.
  THE METHODS OF ETHICS. Third Edition. 8vo. 14s. A Supplement to the Second Edition, containing all the important Additions and Alterations in the Third Edition. Demy 8vo. 6s.
  OUTLINES OF THE HISTORY OF ETHICS, for English Readers. Second Edition, revised. Crown 8vo. 3s. 6d.

Venn.—THE LOGIC OF CHANCE. An Essay on the Foundations and Province of the Theory of Probability, with special Reference to its Logical Bearings and its Application to Moral and Social Science. By JOHN VENN, M.A., Fellow and Lecturer in Moral Sciences in Gonville and Caius College, Cambridge, Examiner in Moral Philosophy in the University of London. Second Edition, rewritten and greatly enlarged. Crown 8vo. 10s. 6d.
  SYMBOLIC LOGIC. By the same Author. Crown 8vo. 10s. 6d.

GEOGRAPHY.

Bartholomew.—THE ELEMENTARY SCHOOL ATLAS. By JOHN BARTHOLOMEW, F.R.G.S. 1s.
  This Elementary Atlas is designed to illustrate the principal textbooks on Elementary Geography.

GEOGRAPHY.

Clarke.—CLASS-BOOK OF GEOGRAPHY. By C. B. Clarke, M.A., F.L.S., F.G.S., F.R.S. New Edition, with Eighteen Coloured Maps. Fcap. 8vo. 3s.

Geikie.—Works by Archibald Geikie, F.R.S., Director-General of the Geological Survey of the United Kingdom, and Director of the Museum of Practical Geology, Jermyn Street, London; formerly Murchison Professor of Geology and Mineralogy in the University of Edinburgh.

THE TEACHING OF GEOGRAPHY. A Practical Handbook for the use of Teachers. Crown 8vo. 2s. Being Volume I. of a New Geographical Series Edited by Archibald Geikie, F.R.S.

*** The aim of this volume is to advocate the claims of geography as an educational discipline of a high order, and to show how these claims may be practically recognised by teachers.

AN ELEMENTARY GEOGRAPHY OF THE BRITISH ISLES. 18mo. 1s.

Green.—A SHORT GEOGRAPHY OF THE BRITISH ISLANDS. By John Richard Green and Alice Stopford Green. With Maps. Fcap. 8vo. 3s. 6d.

Grove.—A PRIMER OF GEOGRAPHY. By Sir George Grove, D.C.L. With Illustrations. 18mo. 1s. (*Science Primers.*)

Hughes.—MAPS AND MAP MAKING. By Alfred Hughes, M.A., late Scholar of Corpus Christi College, Oxford, Assistant Master at Manchester Grammar School. Cr. 8vo. [*In the press.*

Kiepert.—A MANUAL OF ANCIENT GEOGRAPHY. From the German of Dr. H. Kiepert. Crown 8vo. 5s.

Macmillan's Geographical Series. Edited by Archibald Geikie, F.R.S., Director-General of the Geological Survey of the United Kingdom.

The following List of Volumes is contemplated:—

THE TEACHING OF GEOGRAPHY. A Practical Handbook for the use of Teachers. By Archibald Geikie, F.R.S., Director-General of the Geological Survey of the United Kingdom, and Director of the Museum of Practical Geology, Jermyn Street, London; formerly Murchison Professor of Geology and Mineralogy in the University of Edinburgh. Crown 8vo. 2s.

*** The aim of this volume is to advocate the claims of geography as an educational discipline of a high order, and to show how these claims may be practically recognized by teachers.

AN ELEMENTARY GEOGRAPHY OF THE BRITISH ISLES. By Archibald Geikie, F.R.S. 18mo. 1s.

THE ELEMENTARY SCHOOL ATLAS. With 24 Maps in Colours, specially designed to illustrate all Elementary Text-books of Geography. By John Bartholomew. F.R.G.S. 4to. 1s.

MAPS AND MAP MAKING. By Alfred Hughes, M.A., late Scholar of Corpus Christi College, Oxford, Assistant Master at Manchester Grammar School. Crown 8vo. [*In the press.*

AN ELEMENTARY GENERAL GEOGRAPHY. By Hugh Robert Mill., D.Sc. Edin. Crown 8vo. [*In the press.*

A GEOGRAPHY OF THE BRITISH COLONIES.

Macmillan's Geographical Series—*continued.*
A GEOGRAPHY OF EUROPE. By JAMES SIME, M.A.
[*In preparation.*]
A GEOGRAPHY OF AMERICA.
A GEOGRAPHY OF ASIA.
A GEOGRAPHY OF AFRICA.
A GEOGRAPHY OF THE OCEANS AND OCEANIC ISLANDS.
ADVANCED CLASS-BOOK OF THE GEOGRAPHY OF BRITAIN.
GEOGRAPHY OF AUSTRALIA AND NEW ZEALAND.
GEOGRAPHY OF BRITISH NORTH AMERICA.
GEOGRAPHY OF INDIA.
GEOGRAPHY OF THE UNITED STATES
ADVANCED CLASS-BOOK OF THE GEOGRAPHY OF EUROPE.

Mill.—AN ELEMENTARY GENERAL GEOGRAPHY. By HUGH ROBERT MILL, D.Sc. Edin. Crown 8vo. [*In the press.*]

Sime.—A GEOGRAPHY OF EUROPE. By JAMES SIME, M.A. [*In preparation.*]

Strachey.—LECTURES ON GEOGRAPHY. By General R. STRACHEY, R.E., C.S.I., President of the Royal Geographical Society. Crown 8vo. 4s. 6d.

HISTORY.

Arnold (T.).—THE SECOND PUNIC WAR. Being Chapters from THE HISTORY OF ROME. By THOMAS ARNOLD, D.D. Edited, with Notes, by W. T. ARNOLD, M.A. With 8 Maps. Crown 8vo. 8s. 6d.

Arnold (W. T.).—THE ROMAN SYSTEM OF PROVINCIAL ADMINISTRATION TO THE ACCESSION OF CONSTANTINE THE GREAT. By W. T. ARNOLD, M.A. Crown 8vo. 6s.
"Ought to prove a valuable handbook to the student of Roman history."—GUARDIAN.

Beesly.—STORIES FROM THE HISTORY OF ROME. By Mrs. BEESLY. Fcap. 8vo. 2s. 6d.

Bryce.—THE HOLY ROMAN EMPIRE. By JAMES BRYCE, D.C.L., Fellow of Oriel College, and Regius Professor of Civil Law in the University of Oxford. Eighth Edition. Crown 8vo. 7s. 6d.

Buckland.—OUR NATIONAL INSTITUTIONS. A Short Sketch for Schools. By ANNA BUCKLAND. With Glossary. 18mo. 1s.

Buckley.—A HISTORY OF ENGLAND FOR BEGINNERS. By ARABELLA B. BUCKLEY. Author of "A Short History of Natural Science," &c. With Coloured Maps, Chronological and Genealogical Tables. Globe 8vo. 3s.

Bury.—A HISTORY OF THE LATER ROMAN EMPIRE FROM ARCADIUS TO IRENE, A.D. 395-800. By JOHN B. BURY, Fellow of Trinity Coll., Dublin. 2 vols. 8vo. [*In the press.*]

HISTORY.

Dicey.—LECTURES INTRODUCTORY TO THE STUDY OF THE LAW OF THE CONSTITUTION. By A. V. DICEY, B.C.L., of the Inner Temple, Barrister-at-Law; Vinerian Professor of English Law; Fellow of All Souls College, Oxford; Hon. LL.D. Glasgow. Second Edition. Demy 8vo. 12s. 6d.

English Statesmen, Twelve.—A Series of Short Biographies, not designed to be a complete roll of Famous Statesmen, but to present in historic order the lives and work of those leading actors in our affairs who by their direct influence have left an abiding mark on the policy, the institutions, and the position of Great Britain among States.

The following list of subjects is the result of careful selection. The great movements of national history are made to follow one another in a connected course, and the series is intended to form a continuous narrative of English freedom, order, and power. The volumes as follow, Crown 8vo, 2s. 6d. each, are ready or in preparation:—

WILLIAM THE CONQUEROR. By EDWARD A. FREEMAN, D.C.L., LL.D. [Ready.
HENRY II. By Mrs. J. R. GREEN. [Ready.
EDWARD I. By F. YORK POWELL.
HENRY VII. By JAMES GAIRDNER.
CARDINAL WOLSEY. By Professor M. CREIGHTON. [Ready.
ELIZABETH. By the DEAN OF ST. PAUL'S.
OLIVER CROMWELL. By FREDERIC HARRISON. [Ready.
WILLIAM III. By H. D. TRAILL. [Ready
WALPOLE. By JOHN MORLEY. [In the press.
CHATHAM. By JOHN MORLEY.
PITT. By JOHN MORLEY.
PEEL. By J. R. THURSFIELD.

Freeman.—Works by EDWARD A. FREEMAN, D.C.L., LL.D., Regius Professor of Modern History in the University of Oxford, &c.

OLD ENGLISH HISTORY. With Five Coloured Maps. New Edition. Extra fcap. 8vo. 6s.

A SCHOOL HISTORY OF ROME. Crown 8vo. [In preparation.

METHODS OF HISTORICAL STUDY. A Course of Lectures. 8vo. 10s. 6d.

THE CHIEF PERIODS OF EUROPEAN HISTORY. Six Lectures read in the University of Oxford in Trinity Term, 1885. With an Essay on Greek Cities under Roman Rule. 8vo. 10s. 6d.

HISTORICAL ESSAYS. First Series. Fourth Edition. 8vo. 10s. 6d.

Contents:—The Mythical and Romantic Elements in Early English History—The Continuity of English History—The Relations between the Crown of England and Scotland—St. Thomas of Canterbury and his Biographers, &c.

HISTORICAL ESSAYS. Second Series. Second Edition, with additional Essays. 8vo. 10s. 6d.

Contents:—Ancient Greece and Mediæval Italy—Mr. Gladstone's Homer and the Homeric Ages—The Historians of Athens—The Athenian Democracy—Alexander the Great—Greece during the Macedonian Period—Mommsen's History of Rome—Lucius Cornelius Sulla—The Flavian Cæsars, &c., &c.

Freeman—*continued.*
HISTORICAL ESSAYS. Third Series. 8vo. 12s.
Contents:—First Impressions of Rome—The Illyrian Emperors and their Land—Augusta Treverorum—The Goths at Ravenna—Race and Language—The Byzantine Empire—First Impressions of Athens—Mediæval and Modern Greece—The Southern Slaves—Sicilian Cycles—The Normans at Palermo.
THE GROWTH OF THE ENGLISH CONSTITUTION FROM THE EARLIEST TIMES. Fourth Edition. Crown 8vo. 5s.
GENERAL SKETCH OF EUROPEAN HISTORY. New Edition. Enlarged, with Maps, &c. 18mo. 3s. 6d. (Vol. I. of Historical Course for Schools.)
EUROPE. 18mo. 1s. (*History Primers.*)

Fyffe.—A SCHOOL HISTORY OF GREECE. By C. A. FYFFE, M.A. Crown 8vo. [*In preparation.*

Green.— Works by JOHN RICHARD GREEN, M.A., LL.D., late Honorary Fellow of Jesus College, Oxford.
A SHORT HISTORY OF THE ENGLISH PEOPLE. New and Thoroughly Revised Edition. With Coloured Maps, Genealogical Tables, and Chronological Annals. Crown 8vo. 8s. 6d. 136th Thousand.
Also the same in Four Parts. Parts I. and II. ready; Parts III. and IV. shortly. With the corresponding portion of Mr. Tait's "Analysis." Crown 8vo. 3s. 6d. each. Part I. 607—1265. Part II. 1265—1540. Part III. 1540—1660. Part IV. 1660—1873.
HISTORY OF THE ENGLISH PEOPLE. In four vols. 8vo.
Vol. I.—EARLY ENGLAND, 449-1071—Foreign Kings, 1071-1214—The Charter, 1214-1291—The Parliament, 1307-1461. With eight Coloured Maps. 8vo. 16s.
Vol. II.—THE MONARCHY, 1461-1540—The Reformation, 1540-1603. 8vo. 16s.
Vol III.—PURITAN ENGLAND, 1603-1660—The Revolution, 1660-1688. With four Maps. 8vo. 16s.
THE MAKING OF ENGLAND. With Maps. 8vo. 16s.
THE CONQUEST OF ENGLAND. With Maps and Portrait. 8vo. 18s.
ANALYSIS OF ENGLISH HISTORY, based on Green's "Short History of the English People." By C. W. A. TAIT, M.A., Assistant-Master, Clifton College. Crown 8vo. 3s. 6d.
READINGS FROM ENGLISH HISTORY. Selected and Edited by JOHN RICHARD GREEN. Three Parts. Globe 8vo. 1s. 6d. each. I. Hengist to Cressy. II. Cressy to Cromwell. III. Cromwell to Balaklava.

Guest.—LECTURES ON THE HISTORY OF ENGLAND. By M. J. GUEST. With Maps. Crown 8vo. 6s.

Historical Course for Schools—Edited by EDWARD A. FREEMAN, D.C.L., LL.D., late Fellow of Trinity College, Oxford, Regius Professor of Modern History in the University of Oxford.
I.—GENERAL SKETCH OF EUROPEAN HISTORY. By EDWARD A. FREEMAN, D.C.L. New Edition, revised and enlarged, with Chronological Table, Maps, and Index. 18mo. 3s. 6d.
II.—HISTORY OF ENGLAND. By EDITH THOMPSON. New Ed., revised and enlarged, with Coloured Maps. 18mo. 2s. 6d.

HISTORY. 39

Historical Course for Schools—*continued.*
III.—HISTORY OF SCOTLAND. By MARGARET MACARTHUR. New Edition. 18mo. 2s.
IV.—HISTORY OF ITALY. By the Rev. W. HUNT, M.A. New Edition, with Coloured Maps. 18mo. 3s. 6d.
V.—HISTORY OF GERMANY. By J. SIME, M.A. New Edition Revi ed. 18mo. 3s.
VI.—HISTORY OF AMERICA. By JOHN A. DOYLE. With Maps. 18mo. 4s. 6d.
VII.—EUROPEAN COLONIES. By E. J. PAYNE, M.A. With Maps. 18mo. 4s. 6d.
VIII.—FRANCE. By CHARLOTTE M. YONGE. With Maps. 18mo. 3s. 6d.
GREECE. By EDWARD A. FREEMAN, D.C.L. [*In preparation.*
ROME. By EDWARD A. FREEMAN, D.C.L. [*In preparation.*

History Primers—Edited by JOHN RICHARD GREEN, M.A., LL.D., Author of "A Short History of the English People."
ROME. By the Rev. M. CREIGHTON, M.A., Dixie Professor of Ecclesiastical History in the University of Cambridge. With Eleven Maps. 18mo. 1s.
GREECE. By C. A. FYFFE, M.A., Fellow and late Tutor of University College, Oxford. With Five Maps. 18mo. 1s.
EUROPEAN HISTORY. By E. A. FREEMAN, D.C.L., LL.D. With Maps. 18mo. 1s.
GREEK ANTIQUITIES. By the Rev. J. P. MAHAFFY, M.A. Illustrated. 18mo. 1s.
CLASSICAL GEOGRAPHY. By H. F. TOZER, M.A. 18mo. 1s.
GEOGRAPHY. By Sir G. GROVE, D.C.L. Maps. 18mo. 1s.
ROMAN ANTIQUITIES. By Professor WILKINS. Illustrated. 18mo. 1s.
FRANCE. By CHARLOTTE M. YONGE. 18mo. 1s.

Hole.—A GENEALOGICAL STEMMA OF THE KINGS OF ENGLAND AND FRANCE. By the Rev. C. HOLE. On Sheet. 1s.

Jennings—CHRONOLOGICAL TABLES. A synchronistic arrangement of the events of Ancient History (with an Index). By the Rev. ARTHUR C. JENNINGS, Rector of King's Stanley, Gloucestershire, Author of "A Commentary on the Psalms," "Ecclesia Anglicana," "Manual of Church History," &c. 8vo. 5s.

Labberton.—NEW HISTORICAL ATLAS AND GENERAL HISTORY. By R. H. LABBERTON, Litt Hum.D. 4to. New Edition Revised and Enlarged. 15s.

Lethbridge.—A SHORT MANUAL OF THE HISTORY OF INDIA. With an Account of INDIA AS IT IS. The Soil, Climate, and Productions; the People, their Races, Religions, Public Works, and Industries; the Civil Services, and System of Administration. By Sir ROPER LETHBRIDGE, M.A., C.I.E., late Scholar of Exeter College, Oxford, formerly Principal of Kishnaghur College, Bengal, Fellow and sometime Examiner of the Calcutta University. With Maps. Crown 8vo. 5s.

Mahaffy.—GREEK LIFE AND THOUGHT FROM THE AGE OF ALEXANDER TO THE ROMAN CONQUEST. By the Rev. J. P. MAHAFFY, M.A., D.D., Fellow of Trinity Coll., Dublin, Author of "Social Life in Greece from Homer to Menander," "Rambles and Studies in Greece," &c. Crown 8vo. 12s. 6d.

Michelet.—A SUMMARY OF MODERN HISTORY. Translated from the French of M. MICHELET, and continued to the Present Time, by M. C. M. SIMPSON. Globe 8vo. 4s. 6d.

Norgate.—ENGLAND UNDER THE ANGEVIN KINGS. By KATE NORGATE. With Maps and Plans. 2 vols. 8vo. 32s.

Otté.—SCANDINAVIAN HISTORY. By E. C. OTTÉ. With Maps. Globe 8vo. 6s.

Ramsay.—A SCHOOL HISTORY OF ROME. By G. G. RAMSAY, M.A., Professor of Humanity in the University of Glasgow. With Maps. Crown 8vo. [*In preparation.*

Seeley.—Works by J. R. SEELEY, M.A., Regius Professor of Modern History in the University of Cambridge.
  THE EXPANSION OF ENGLAND. Crown 8vo. 4s. 6d.
  OUR COLONIAL EXPANSION. Extracts from the above. Crown 8vo. Sewed. 1s.

Tait.—ANALYSIS OF ENGLISH HISTORY, based on Green's "Short History of the English People." By C. W. A. TAIT, M.A., Assistant-Master, Clifton College. Crown 8vo. 3s. 6d.

Wheeler.—A SHORT HISTORY OF INDIA AND OF THE FRONTIER STATES OF AFGHANISTAN, NEPAUL, AND BURMA. By J. TALBOYS WHEELER. With Maps. Crown 8vo. 12s.
  COLLEGE HISTORY OF INDIA, ASIATIC AND EUROPEAN By the same. With Maps. Crown 8vo. 3s. 6d.

Yonge (Charlotte M.).—CAMEOS FROM ENGLISH HISTORY. By CHARLOTTE M. YONGE, Author of "The Heir of Redclyffe," Extra fcap. 8vo. New Edition. 5s. each. (1) FROM ROLLO TO EDWARD II. (2) THE WARS IN FRANCE. (3) THE WARS OF THE ROSES. (4) REFORMATION TIMES. (5) ENGLAND AND SPAIN. (6) FORTY YEARS OF STUART RULE (1603—1643).
  EUROPEAN HISTORY. Narrated in a Series of Historical Selections from the Best Authorities. Edited and arranged by E. M. SEWELL and C. M. YONGE. First Series, 1003—1154. New Edition. Crown 8vo. 6s. Second Series, 1088—1228. New Edition. Crown 8vo. 6s.
  THE VICTORIAN HALF CENTURY—A JUBILEE BOOK. With a New Portrait of the Queen. Crown 8vo., paper covers, 1s. Cloth, 1s. 6d.

www.ingramcontent.com/pod-product-compliance
Lightning Source LLC
Chambersburg PA
CBHW030323240426
43673CB00040B/1253